NANOSCALE

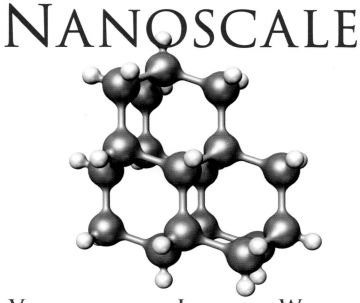

VISUALIZING AN INVISIBLE WORLD

Words by
Kenneth S. Deffeyes

Illustrations by
Stephen E. Deffeyes

The MIT Press
Cambridge, Massachusetts
London, England

MIT Press books may be purchased at special quantity discounts for business or sales promotional use. For information, please email special_sales@mitpress.mit .edu or write to Special Sales Department, The MIT Press, 55 Hayward Street, Cambridge, MA 02142.

Printed and bound in Spain.

Library of Congress Cataloging-in-Publication Data

Deffeyes, Kenneth S.
Nanoscale : visualizing an invisible world / Kenneth S. Deffeyes and Stephen E. Deffeyes.
    p.   cm.
Includes bibliographical references.
ISBN 978-0-262-01283-6 (hardcover : alk. paper)
1. Science.   2. Molecular structure.   3. Nanostructures.   4. Crystallography.
5. Nanoscience.   I. Deffeyes, Stephen E.   II. Title.
Q158.5D454   2009
541'.24—dc22

                                                                    2008032279

10   9   8   7   6   5   4   3   2   1

Dedicated to the memory of

Linus Pauling and Roger Hayward

CONTENTS

INTRODUCTION

Before 1900, virtually all science concerned things that you could see. Telescopes and microscopes extended our vision, but before Einstein's 1905 papers on fluid properties (including Brownian motion), belief in the existence of atoms and molecules was optional; if they did exist we knew only that they were too small to see with a visible-light microscope. Einstein established the size, and therefore the reality, of atoms and molecules (see p. 130).

How can we "see" the atomic structure of our world? Most atoms are roughly 0.2 nanometers in diameter, where a nanometer is a billionth of a meter. Even a scanning electron microscope can give us only details 1 to 5 nanometers in size. But no microscope, however powerful, could take a picture of the way atoms actually look. At that scale, the laws of quantum mechanics apply, and the electrons that make up the outer parts of atoms consist only of a diffusion of probability. In illustrating this book, we have used various graphic conventions to translate that fuzzy world into our ordinary-scale experience. Through these illustrations we can visualize the intricate dance of forces that make up our seemingly solid world.

In one sense, this book is an update of *The Architecture of Molecules*, published in 1964 by Linus Pauling and Roger Hayward. We say that with some trepidation – Pauling was one of the greatest scientists of his time and Hayward, an architect, was an enormously talented scientific illustrator. Like Pauling and Hayward's book, this book does not make an attempt at exhaustive coverage. Each of the 50 subjects here was selected because it illustrates how atomic structure creates a property such as hardness, color, or toxicity; because it has a great story; or sometimes simply because it is beautiful.

Crystallography, the science of determining the arrangement of atoms in solids, is important in physics, chemistry, geology, biology, metallurgy, and engineering. However, only a tiny number of universities have had departments of crystallography. We counted more than a dozen Nobel Prize winners who advanced crystallography while hiding out in other departments.

Progress in understanding the structure of crystals was initially limited by a lack of suitable tools. In 1914, Max von Laue discovered that x-rays behaved as waves that could be diffracted by the atoms within crystals. The diffracted rays reflect the symmetry and structure of the crystal. After 1945, x-ray diffraction devices became widely available, and since then the atomic structure of thousands of substances has been determined.

Stories from the history of crystallography and chemistry can be found scattered throughout the book. Here is a brief timeline of the major developments:

1669 ❋ Nicholas Steno recognized the constancy of angles between crystal faces.

1800 ❋ Rene Just Haüy showed that crystal faces reflect underlying building blocks.

1854 ❋ James Dwight Dana organized natural minerals into chemical families.

1891 ❋ Both Federov and Schoenflies identified the 230 possible crystal symmetries.

1905 ❋ Albert Einstein showed the reality, and the size, of atoms.

1912 ❋ Max von Laue showed that crystals diffract x-rays in an orderly way.

1912 ❋ Bragg and Bragg, father and son, started studying crystals and x-rays.

1923 ❋ Linus Pauling started determining chemical principles from crystal structure.

1935 ❋ Kathleen Lonsdale published the first *International Tables for Crystallography*.

1957 ❋ Watson and Crick published the structure of DNA.

1984 ❋ Schechtman discovered quasicrystals.

Readers of this book may recognize *crystal* as something shiny on a neck chain, *minerals* as ingredients in vitamin pills, and *rock* as loud music. For our purposes, a crystal is a solid composed of a repeating array of atoms. A mineral is a natural solid having a definite crystal structure. Geologists have always had a penchant for naming things, often quite poetically, and even if several minerals have the same chemical composition, each of the crystal structures gets its own name. A rock is a natural solid composed of one or more minerals.

On the chemical side, an *atom* is the smallest unit identifiable in the chemical elements numbered from 1 (hydrogen) to 92 (uranium.) An atom that has an electrical charge from gaining or losing one or more electrons is called an *ion*.

In our illustrations, we have consistently shown oxygen atoms as red, nitrogen as blue, carbon as black, and hydrogen as white. Colors for other atoms will be identified in the individual essays.

We hope you enjoy this tour through the nano-scale world. Stephen prepared the illustrations and Kenneth wrote the essays. The selection of subjects, the identification of sources, and verifying the information has been a joint effort.

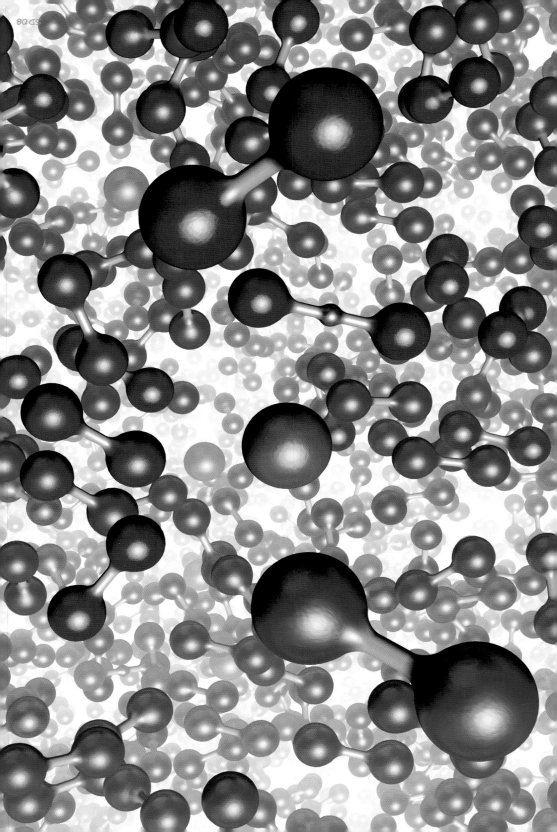

1   AIR

Air is a mixture of several kinds of molecules. Nitrogen atoms, shown in blue, combine in pairs and form 77 percent of the air. Oxygen atoms, the red pairs, amount to 22 percent. Argon (violet), which does not form chemical bonds, is shown as single atoms that make up most of the remaining 1 percent. Carbon dioxide, despite its role in Earth's climate, amounts to only 0.03 percent. Carbon dioxide is identifiable in the illustration as two oxygen atoms linked by a small black carbon atom.

Although we all think of atmospheric oxygen as A Good Thing, there is an alternative way of thinking about it. Earth's early atmosphere contained almost no oxygen. Oxygen is – in that sense – a pollutant, a vile poison, a toxic byproduct of green-plant photosynthesis. As oxygen accumulated, multi-celled animals that used the free oxygen to drive their metabolism emerged. However, oxygen is allowed in our bodies only when attached to chemical escorts: myoglobin and hemoglobin (#16).

There is a continuing discussion about the time when free atmospheric oxygen appeared. Part of the confusion arises because some local environments may have become oxygenated long before the main atmosphere accumulated significant oxygen.

The present-day atmospheric level of 22 percent oxygen is not "engraved in stone." Green plants take up carbon dioxide from the air and liberate oxygen. After the plant dies, bacteria use oxygen from the air or from seawater to convert the plant material back to carbon dioxide. However, if the plant material is buried to form a coal bed the oxygen is stranded in the atmosphere. Bob Berner at Yale has developed convincing evidence that 300 million years ago, when the Appalachian coal beds were being deposited, atmospheric oxygen rose to twice its present-day level. A forest fire would have been a catastrophic event.

Previous Page:
Major components of Earth's atmosphere.

2   ICE AND WATER VAPOR

The lower part of the illustration shows water molecules as they occur in water vapor (steam). Each oxygen (red) is accompanied by two hydrogens (white); with the two hydrogens separated by an angle of 105 degrees. The arrangement is irregular, both in the spacing between molecules and in their orientation.

At the top is the structure of crystalline ice, as first determined in 1935 by Linus Pauling. Each oxygen has four nearest neighbors and the angle between nearest neighbors is 109.5 degrees. It takes only a minor distortion to fit the preferred 105 degrees in the isolated water molecule into the ice structure. The hydrogens are not located midway between the oxygens. Each oxygen has two nearby hydrogens, maintaining the familiar formula H_2O. The gray box shows that the ice structure is hexagonal; which accounts for the six-sided symmetry of snowflakes.

Liquid water probably contains temporary local domains that have the ice structure, but there is no long-range ordering in distance or direction that characterizes a typical crystal. A close-up view of liquid water would locally resemble the hexagonal ice structure.

Geologists think of ice as a mineral. A mineral is a naturally occurring substance with a definite crystal structure. Ice fits. On mornings before geology field trips the parking lot sometimes contains sedimentary ice (undisturbed fresh snow), igneous ice (frozen puddles), and metamorphic ice (recrystallized beneath tire tracks).

Ice and water are unusual in several ways. Ice cubes float on water. However, for most substances the solid is denser than the liquid. Only a half-dozen materials expand while freezing or solidifying. One historically important example was type metal, a lead-tin-bismuth alloy used to cast moveable printing type. The molten alloy expanded when it solidified and made an accurate cast from the mold. Silicon, used in semiconductors, is another of the half-dozen materials that expand while freezing.

Ice is also unusual in having substantial strength even when it is one degree colder than its melting point. Most substances, including steel, have

Previous Page:
   Water vapor molecules, surrounding the hexagonal unit cell of ice.

very little strength when heated close to their melting points. We know, we know, you can crush ice cubes with your teeth. One dentist said that ice cubes and unpopped popcorn kernels kept dentists in business.

The hexagonal ice structure, called *Ice Ih*, is stable at room pressure just below its freezing point. However, at other pressures and temperatures ice crystallizes in a dozen different crystal structures. Many of the alternative ice structures were determined by Linus Pauling's son-in-law, Barclay Kamb. (Kamb said that he never discussed crystallography with his father-in-law, because Pauling could solve the problems so fast that it wasn't even fun.)

One ice polymorph, Ice IX (or ice-nine) was made famous in Kurt Vonnegut's 1963 novel *Cat's Cradle*. The fiction was that Ice IX was more stable than ordinary ice (or water) and a chunk of Ice IX would grow to freeze the entire ocean. Vonnegut's brother, Bernard, was a meteorologist who discovered that silver iodide had a crystallographic unit cell very close to ordinary Ice Ih. Modified fireworks that produced nanoscale silver iodide particles could generate ice from undercooled air. As the ice crystals grew, large amounts of heat were liberated by the conversion of water vapor to ice. A dollar's worth of silver iodide, sprayed into moisture-laden air, can release more energy than a hydrogen bomb.

The beautiful symmetry of snowflakes is a consequence of the hexagonal structure of ordinary Ice Ih. However, be warned that the pretty snowflake pictures in books, including this one, are carefully selected. Most real snow-flakes are hexagonal, but not that perfect.

Above:
  A couple of somewhat idealized hexagonal snowflakes.

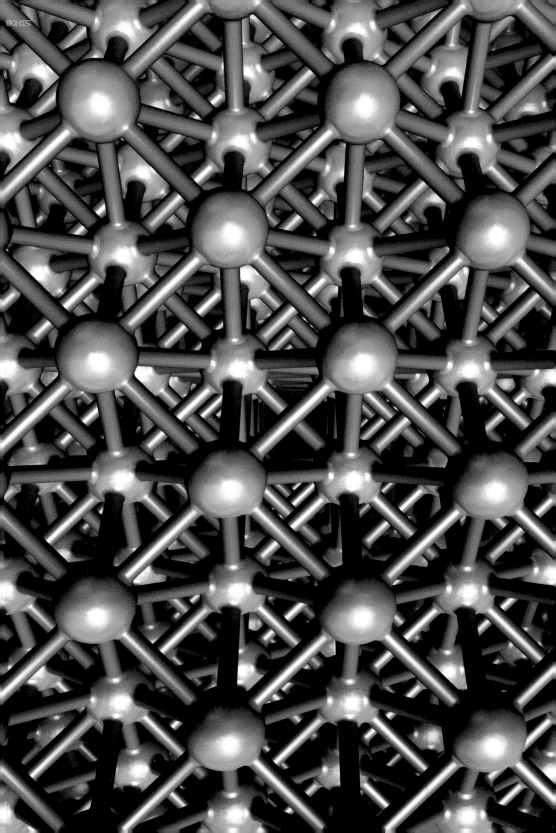

3 GOLD

Pure gold consists of gold atoms packed neatly together in a cubic array. The sticks between atoms show the connections between nearest neighbors. Pure gold is soft and easily melted. (Metallic bonding is discussed in the next section.) Also, because gold sometimes occurs as gold grains in river gravels, it can be mined with simple tools. As a result, we find beautiful gold jewelry thousands of years old. Gold was the first metallic resource to be mined and used.

Gold has two major uses: as a material and as a form of money. Because it does not corrode, gold is useful for jewelry and for electronic contacts. Since mining gold has almost always involved a lot of labor and energy, gold became a standard measure of value.

We usually think of gold mining as something like the California gold rush of 1849. Around 1900, underground gold mining largely replaced pulling gold nuggets out of river gravels. Something almost as exciting as the California gold rush happened in the last half of the twentieth century, and it also happened in the United States. In 1956, Newmont, a mining company, opened a new gold mine near Carlin, Nevada. The ore did not look like conventional gold ore; one geologist said the ore looked like a throwing rock from the driveway. Further, instead of the conventional association of gold with copper, lead, and zinc, the Carlin ore contained arsenic, antimony, mercury, and thallium. Thallium? Who ever heard of thallium? The major reason the Carlin-type gold deposits went undiscovered for so long was its nano-sized gold grains. The old-time prospectors couldn't see it in their gold pans.

At first, the search in Nevada was focused on places with geology similar to Carlin. But then it branched out and Carlin-type gold deposits were found in several different geologic settings. Around the year 2000, Nevada's gold production was exceeded only by that in South Africa and Australia. In 2001, Nevada produced 253 tons of gold metal.

Previous Page:
    The cubic crystal lattice of gold.

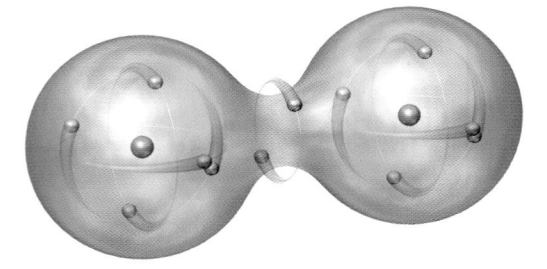

Carbon to Carbon Covalent Bonding

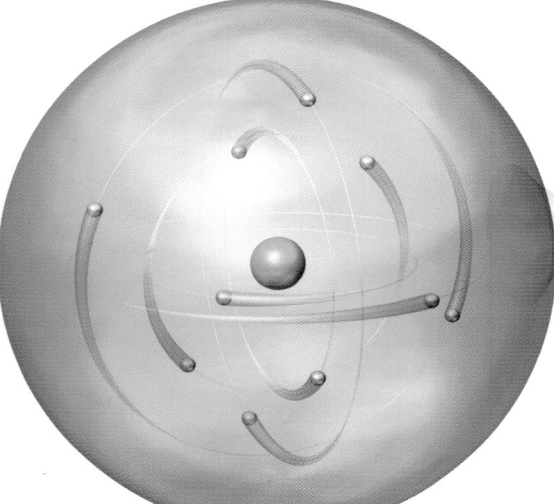

Chlorine to Sodium Ionic Bonding

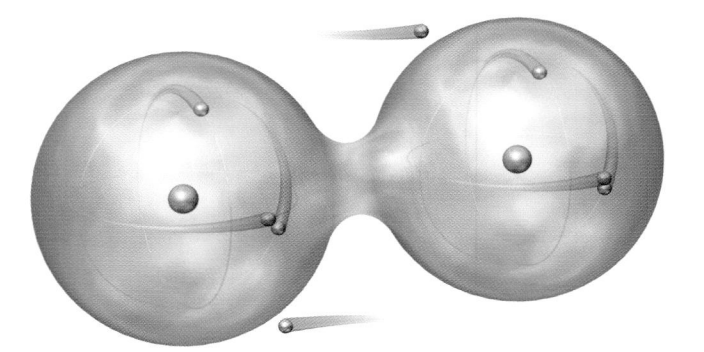

Beryllium to Beryllium Metallic Bonding

SD 08

4   CHEMICAL BONDS

Linus Pauling's 1954 Nobel Prize was for explaining "the nature of the chemical bond." He showed that there were several, quite different, kinds of chemical bonding. Three important types of bonding arise from the outermost electrons of the atoms. From the top down in the illustration, we have:

* Covalent bonding, with the electrons held tightly between the parent atoms. The example shown is an adjacent pair of carbon atoms, as in the structure of diamond (#6). Diamond is the hardest known substance; covalent bonding is typically very strong.

* Ionic bonding, with one of the parent atoms having custody of the electrons. (You are free to assign your own analogies to these examples.) Ordinary table salt, sodium chloride, is the example. Atoms are most stable if they have all their electron shells filled neatly to the rim. Chlorine has an outer electron shell that needs one more electron to fill it. Sodium has one extra electron outside of its inner complete shells. The results, after the transfer, are called sodium ions and chloride ions.

* Metallic bonding, with some of the electrons wandering out in the street, unsupervised. Beryllium metal is the example. Free electrons give metals their characteristic signature: they are good conductors of electricity.

In 1932, Pauling generated a numerical scale of electronegativity, the drive for atoms of a particular element to acquire electrons. (It isn't *electropositivity* because positive and negative electrical voltages were named long before electrons were found to have negative charges.) Any two elements with very different electronegativity will form ionic bonds. Two identical atoms, like two carbons or two berylliums, have to form either metallic or covalent bonds. A count of the chemical elements showed 51 metals and 16 covalent nonmetals.

Previous Page:
    Covalent, ionic, and metallic chemical bonds, showing the
    distribution of outer electrons.

5   SODIUM CHLORIDE

After the diffraction of x-rays by crystals was discovered in 1912, the first crystal structure to be deciphered was sodium chloride, table salt. Most chemists expected that molecules containing one sodium atom attached to one chlorine atom would make up the structure. No such luck. Molecules were nowhere to be seen. Instead there was a three-dimensional checkerboard of sodium ions (shown in white on the facing page) alternating with chloride ions (shown in green). In the illustration, sodium ions and chloride ions are shown dissolved in salt brine surrounding the crystal. Again, molecules of sodium chloride are not present.

The structure of sodium chloride was discovered by a team of father (William Henry Bragg) and son (William Lawrence Bragg). The son used x-rays to study the structure of crystals and the father used crystals to study the properties of x-rays. The Braggs made use of nearly perfect natural crystals, inches across, of sodium chloride (#5) and calcite (#22). The two Braggs shared the Nobel Prize for Physics in 1915.

Since ancient times, salt has been used for preserving meat and fish. Salting meat is the chemical equivalent of drying; both make water unavailable to microorganisms. *Salary* (worth his salt) and *salami* come from the Latin word for salt.

Salt was obtained by evaporating sea water, or by mining, as it still is today. The natural evaporation of ancient sea water and lake water formed extensive geologic layers of sodium chloride and other chemicals, such as potassium chloride. Potassium chloride deposits are mined today as the major source of potassium in fertilizer. After one of my blood tests turned out to be a bit low in potassium, my physician prescribed daily pills of potassium chloride. I figured out that my pills cost the mine-mouth price of potassium chloride, multiplied by 30,000.

Previous Page:
    A crystal of sodium chloride (table salt), surrounded by the sodium and chloride ions found in salt brine.

6   DIAMOND

The crystal structure of diamond was investigated early, in the same year as sodium chloride. Each carbon atom has four covalent bonds attached to its nearest neighbors in a cubic array. Because of the strength of the covalent bonds, diamond is the hardest known substance. In a sense, a diamond crystal is one big molecule. Synthetic diamonds, as well as natural diamonds that are not clear enough to use as gemstones, play an important industrial role as abrasives and as cutting tools. For instance, oil-well drill bits faced with diamond have cut wells as deep as 22,000 feet without the need to pull the drill pipe out to change bits.

Diamonds are enormously valuable, but all the world's primary diamond mines put together would add up to an area of 20 square kilometers, compared to Earth's land surface of 150 million square kilometers. Discovery of a new diamond mine is a major thrill. My favorite yarn is the 1940 discovery by John Williamson (a Canadian with a Ph.D. in geology) of the Mwadui mine in what was then Tanganyika. Williamson knew that a few stray diamonds had been found in east Africa. He was searching by himself and out of money. His car got stuck in a mudhole, he left the car turning over in low gear, and he got out to push. He reached down and pulled a diamond out of the mud. A mine was opened and, in 1947, Williamson presented Princess Elizabeth (later Queen Elizabeth II) with a flawless 23-carat pink diamond as a wedding present.

Before 2007, the world market for gem-quality diamonds was controlled by the De Beers company. De Beers was founded in 1888 by Cecil Rhodes, who later endowed the Rhodes Scholarships. De Beers could not operate in the United States because it was a cartel, guilty of price fixing and "a conspiracy in restraint of trade." That did not keep De Beers from running ads in the United States, including advice that the proper price for a diamond engagement ring was two months' salary. In order to maintain diamond prices during difficult economic times, De Beers would hold back as much as a billion dollars worth of uncut diamonds. Gradually, around 2007, De Beers ceased trying to set diamond prices; they have now opened a store on Fifth Avenue in New York.

Previous Page:
The cubic crystal lattice of diamond.

We have given the diamond structure a greasy look, because natural and synthetic crystalline diamonds are inherently water-repellant. Most diamonds are recovered by running the crushed ore or natural gravel across grease-coated tables. Occasionally, you scrape the grease off the table, run the melted grease through a sieve, and there are your diamonds. A do-it-yourself test for a genuine diamond is to clean the stone with detergent and thoroughly rinse it. Then, with the sharpened end of a toothpick, place a drop of water on top of the diamond. If the water drop does not spread out, the surface is greasy and it is probably a genuine diamond.

Diamonds are stable at the temperature and pressure typical of depths more than 200 kilometers (125 miles) below Earth's surface. If the diamonds were raised slowly to the surface, they would recrystallize as graphite (#32); they would wind up in pencils instead of engagement rings. Diamonds survive the trip only if they come up quickly. Explosive volcanoes seem to be the only elevator that runs fast enough, and diamonds are usually found along "hotspot" tracks left by ancient volcanoes.

The most recent discoveries of gem-quality diamonds have been in Canada. For more than 100 years, occasional diamonds have been found, in the northern United States and southern Canada, in glacial debris pushed south by continental glaciers. Jason Morgan, who discovered plate tectonics, had mapped hotspot tracks worldwide, including some across northern Canada. Aha! Follow the direction of glacial scratches on the bedrock northward from the stray diamonds until you cross one of Morgan's hotspot tracks. X marks the spot. Two Canadian diamond mines are open and three more mines are coming online.

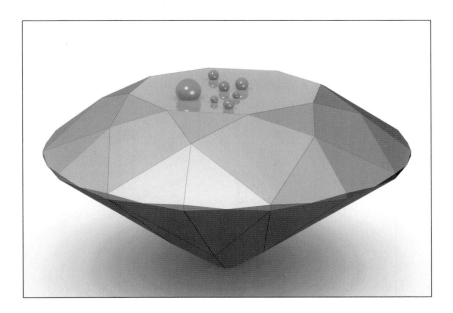

Above:
  Water droplets beaded up on an inherently greasy diamond.

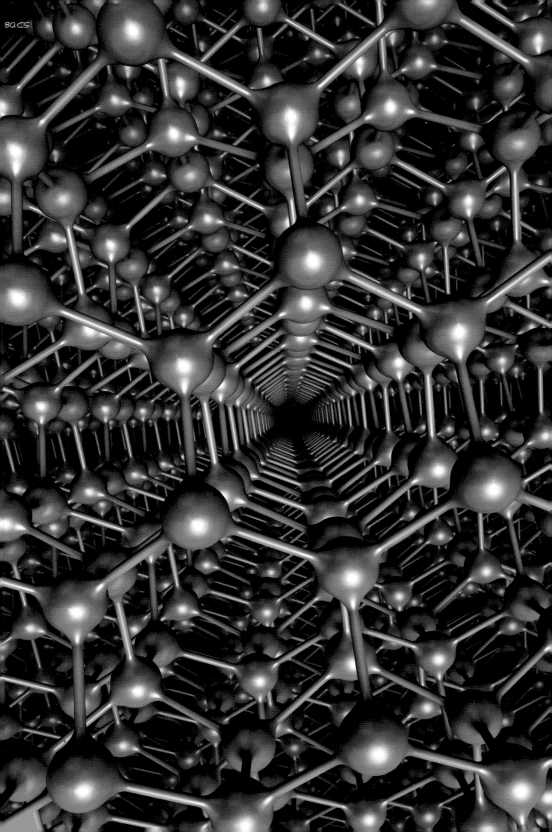

7   HEXAGONAL DIAMOND

A structure quite similar to natural cubic diamond packs the carbon atoms in a hexagonal array instead of a cubic arrangement. In nature, the hexagonal carbon arrangement has been found in the debris from large meteorite impacts, but nowhere else.

A number of other crystals, natural and synthetic, can exist in either hexagonal or cubic form. One notable example is zinc sulfide, which is the cubic mineral sphalerite and the hexagonal mineral wurtzite. Some sphalerite crystals have stepped faces where the crystals grew the wurzite structure for a while then jumped back to the cubic sphalerite arrangement. The distinction between cubic and hexagonal metals is explained later in the book (#28).

Synthetic hexagonal diamond has been produced. It remains to be seen whether a synthetic mixture of hexagonal and cubic diamond would be more successful as an abrasive than either structure alone.

The hexagonal diamond has been named *lonsdaleite* in honor of Kathleen Lonsdale. She was a pioneering investigator in both theory and laboratory crystallography, and she assembled the invaluable *International Tables for Crystallography*. Crystallography may be unique among the sciences in having a substantial number of outstanding women, even 80 years ago.

Despite her fame as a scientist, Kathleen Lonsdale (a Quaker) was sentenced to a prison term during World War II for her refusal to cooperate with war preparations in England. After she was released from prison, she was appointed to the board of governors of the prison and the British government later awarded her the lady equivalent of a knighthood.

Previous Page:
The hexagonal structure of lonsdaleite.

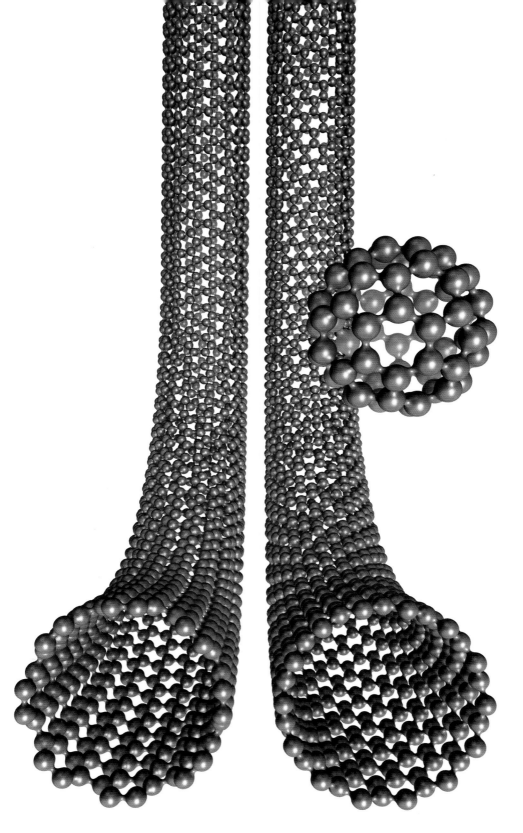

8   NANOTUBES AND BUCKYBALLS

Two forms of carbon, diamond and graphite, have been used since ancient times. It was not until 1985 that the ball-shaped and tubular forms of carbon were produced. Rick Smalley shared the 1996 Nobel Prize for Chemistry for the discovery of the ball-shaped carbon cage. Smalley and his colleagues named the balls *buckminsterfullerene* in honor of R. Buckminster Fuller, who had used the same geometry for his geodesic domes. The name promptly got shortened to *buckyball.*

Smalley realized that the electrically conducting carbon nanotubes could be used to build lightweight electric power transmission lines. The power lines could be used to transport electricity from areas with abundant sunshine (Arizona) or strong winds (Wyoming) to the population centers in the U.S. northeast. Smalley also pointed out that electrical storage could be local. On my grandparents' farm, wind-generated electricity was stored in lead-acid batteries in the cellar. Today, carbon nanotubes are being evaluated as a component for electrical storage in supercapacitors (#39).

Existing manufacturing processes tend to make a complex mixture of all sorts of nanotubes. Until his death from leukemia in 2005, Smalley was spending long hours in his labs at Rice University, searching for a selective process that would produce the electrically conductive nanotubes. Rick Smalley is sorely missed in the alternative-energy community, both because of his chemistry skills and for his vision.

The carbon nanotube on the right is a better electrical conductor than copper. The tube on the left is, at best, an electrical semiconductor. Can you spot the difference? (The answer is at the bottom of this page.)

The electrically conducting nanotube on the right has one set of carbon-carbon bonds in the hexagonal carbon lattice running around the tube. The tube on the left has lines oriented along the length of the tube.

Previous Page:
    Carbon structures: a buckyball and two different nanotubes.

9   ASBESTOS

Two different minerals have been used commercially as asbestos. In the illustration, the structure on the left is a particular member of the amphibole family that separates easily into thin fibers. On the right is a member of the serpentine family (chrysotile) whose nano-structure is like a rolled-up carpet. Plenty of other minerals would have good heat resistance and insulating properties, but they do not separate into fibers.

Beginning in the 1970s, asbestos has been regulated as a hazardous substance. If the two varieties of commercial asbestos are such different minerals, would we expect both of them to be equally dangerous? Malcolm Ross – a distinguished mineralogist, now retired from the U.S. Geological Survey – has long insisted that amphibole asbestos is highly toxic and the serpentine asbestos is either very mildly toxic or not a hazard at all. However, current safety standards refer only to mineral "fibers" and make no distinction between the amphibole and serpentine varieties of asbestos.

Almost all of the amphibole asbestos was mined in South Africa. Those mines are now closed because the material is now recognized as too toxic to process and use. The South African deposits occur in a *banded iron formation*, sometimes called BIF. (I tried to generate an adjective *biffy* but nobody liked it.) Of the world's extensive BIF deposits, only the ones in South Africa, and possibly Australia, contain fibrous asbestos.

Banded iron formations are not forming today; all of them are older than 1.8 billion years. Most of the world's iron ore comes from these old BIF deposits. Their banding is a centimeter-scale layering, with two or more of: iron oxide, iron carbonate, iron sulfide, and iron silicate. Possibly, the chemistry of the early ocean and atmosphere allowed iron to travel as a dissolved chemical species, like the sodium, potassium, calcium, and magnesium in today's ocean.

Serpentine minerals are themselves interesting. A rock made of serpentine minerals is supposed to be called a *serpentinite*, but virtually all geologists use *serpentine* to refer both to the mineral and to the rock. Serpentine has the same chemical composition as Earth's mantle plus water. Just add water. That

Previous Page:
(Right) a serpentine asbestos structure similar to a rolled-up carpet and (left) an amphibole crystal splitting into asbestos fibers.

process is happening today. I found a polished partially serpentinized rock in Westminster Abbey that had raised bumps where the last of the pre-existing rock was actively being converted to serpentine in the wet British climate.

All of the serpentine minerals are made of double layers: one layer of silicon and oxygen and the other layer of magnesium and oxygen. It happens that the silicon-oxygen layers are a bit smaller than the magnesium-oxygen layers, so the layers try to curve and the carpet roll is one way of accommodating the curve.

Serpentinite is the state rock of California. The photograph on the next page shows the most photographed serpentine outcrop in California, although it contains no asbestos. Serpentine is green and feels slick. (That's how it got the "serpent" name.) The rock at the south end of the Golden Gate Bridge was so slick that the ends of the cables holding up the bridge are anchored into an enormous concrete block. The south tower was originally planned to go straight down into the water, but during the construction the base of the tower was enlarged six times. The boat-shaped barrier around the base of the south tower is the outline of a large concrete pad that distributes the load over the weak serpentine beneath. If you have any doubts about this story, look at the slope on the right-hand side of the photograph. It's a mass of separate landslides.

Next Page, Top Left:
 The fundamental building block of amphibole asbestos is highlighted by the darker atoms in the center. Splitting along the weakest bonds generates a particularly toxic form of asbestos.

Next Page, Top Right:
 End view of a serpentine asbestos fiber, with the silicon-oxygen layers on the inside of the carpet roll.

Next Page, Bottom:
 The base of the south tower of the Golden Gate Bridge and the cable anchorage are influenced by the weak serpentine rock beneath.

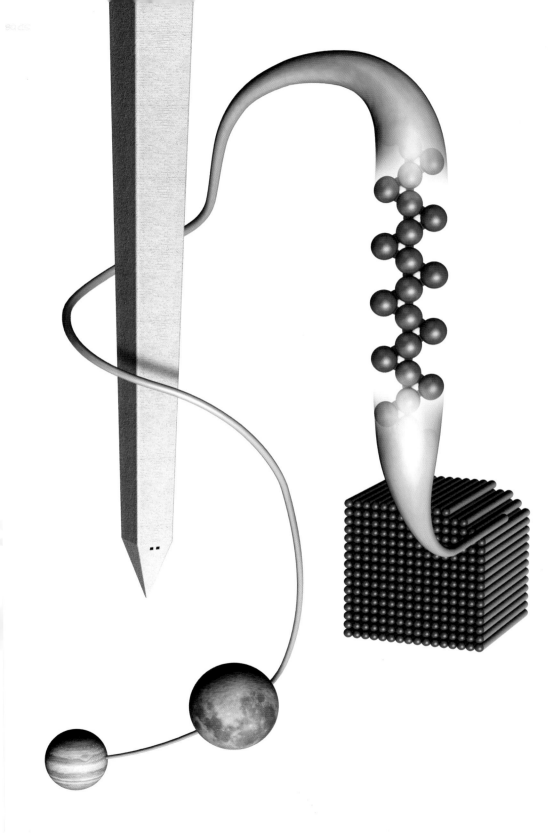

10   PYROXENE

Crystals of the common rock-forming mineral pyroxene are made up of chains of silicon and oxygen atoms, spaced a nanometer (10^{-9} meter or 10^{-7} centimeter) apart. I gave this as an exam question:

> Unravel a cube of pyroxene, one centimeter (0.4 inch) on a side into a single chain. Would the chain reach from the White House to:
>
> __ The Washington Monument (one kilometer, 10^5 cm)
>
> __ Chicago (1000 kilometers)
>
> __ The Moon (300,000 kilometers)
>
> __ Jupiter (1,000,000,000 kilometers)

The correct answer is Jupiter. It's a dirty exam question because most students do it right but then can't believe their answer. If the chains on the end of the crystal are 10^{-7} centimeters apart, then there are 10^7 chains along each edge of the one-centimeter crystal; 10^{14} chains in all. Dividing by 10^5, to convert from centimeters to kilometers, gives 10^9 kilometers, the distance to Jupiter. You can do it in your head: $7 + 7 - 5 = 9$.

The silicon and oxygen atoms lose or gain electrons from each other in order to make complete electron shells; they become electrically charged ions. One of Linus Pauling's great contributions was a set of rules for packing spherical ions together in a way that makes the overall structure electrically neutral on the smallest possible scale.

The basic building block of most of the common rock-forming minerals is a silicon ion surrounded by four oxygen ions. In the commercial plastics industry, *polymers* are plastics built up from simple organic molecules. Minerals show different degrees of polymerization. Olivine contains individual silicon-oxygen groups; it would be a *monomer* in a plastics factory. Combining the groups into chains makes pyroxene. Double chains are amphiboles (#9), sheets are mica (#29), and three-dimensional frameworks include quartz, feldspar, and zeolite (#27, #46, and #31).

Curiously, pyroxene seems to be a name without an official pronunciation. Around geology departments PEER ox ene, pie ROCK scene, PERKS ene, and other variations are heard.

Previous Page:
A one centimeter (0.4 inch) cube of pyroxene unraveled would reach to. . . .

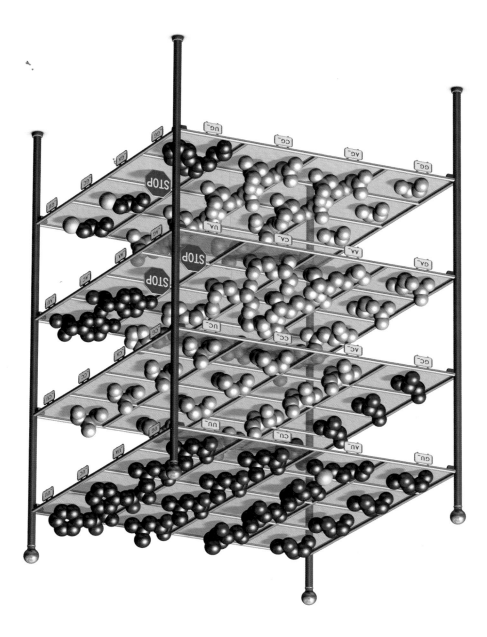

11   AMINO ACIDS

In 1944, Edwin Schroedinger published a short book called *What Is Life?* He observed that something had to carry biological "building instructions" from generation to generation. To make the information stable and reliable, bonds at least as strong as the chemical bonds between atoms would be needed.

Schroedinger speculated that something like an "aperiodic crystal" would be needed to carry the information. (A periodic structure, one that repeats a simple motif over and over, does not carry much information.) Schroedinger was famous for his contributions to quantum mechanics and his book was widely read.

The DNA structure, developed by James Watson and Francis Crick in 1953, turned out to be an aperiodic chain rather than an aperiodic crystal. Over the next 20 years scientists gradually discovered that DNA was transcribed (through RNA as an intermediate) into various chains of amino acids. The resulting long floppy chain of amino acids is not of much biological use, but wadding each type of amino acid chain up into a ball with a tightly defined geometry produces globular proteins. Proteins do most of the work of building and managing cells and protein folding is a topic of great current interest in biology. The first stage of understanding came in 1959 when a chemist, Walter Kauzmann, pointed out that about half of the amino acids were hydrophobic – they were oily and hated water. The others were hydrophilic, water lovers. In a water-based biology, Kauzmann inferred that the hydrophilic amino acids would be on the outside of a globular protein, with the hydrophobic ones curled up in the middle.

The illustration shows which amino acid is specified by each "three letter" sequence of the molecules in DNA and RNA. The hydrophobic amino acids are dark, the hydrophilic ones are light. Observe that changing any one letter usually substitutes the same or a similar amino acid. This means that most errors in transcribing DNA are not fatal. Also, it suggests that a lot of natural selection went on before the present process was standardized. The STOP signs are signals to quit building and to release the amino acid chains.

Previous Page:
Amino acid molecules arranged by the three-base code in DNA and RNA; water-loving amino acids are shown lighter, water-repellent amino acids are darker.

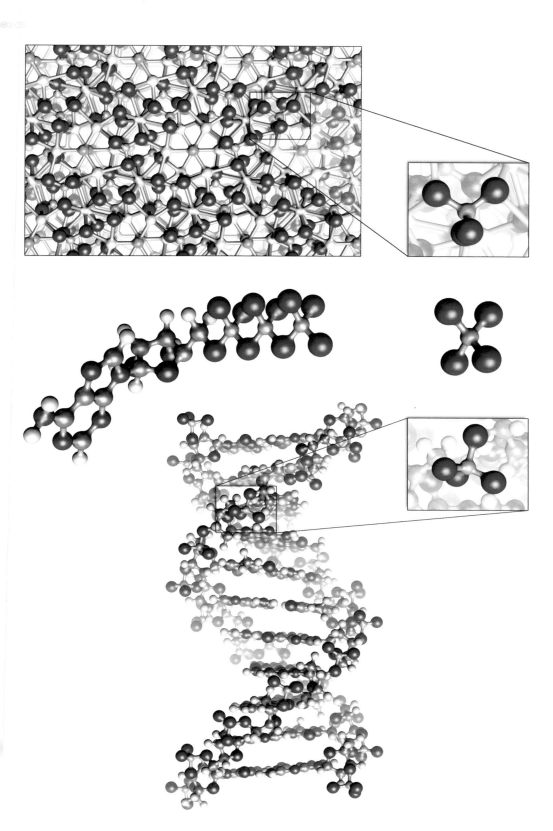

12   PHOSPHATE

The use of the chemical element phosphorous in several vital biological roles comes as a surprise. In virtually all natural environments, phosphorous is tightly bound to four surrounding oxygen ions to form a unit known as *phosphate*. Phosphate is soluble in seawater (and other natural waters) at the level of only a few parts per million. Why then does it turn up as an important part of biological processes? Was it an accident of history or is there no other way to manage a water-based biology?

The illustration shows three major biological roles for phosphate, although there are more.

  * The famous DNA genetic molecule has a row of phosphates on the outside of the helix. These are highlighted in the top illustration with phosphorous in yellow and oxygen in red.
  * The ATP molecule (adenosine triphosphate) is the carrier of energy in all living cells, as shown in the middle picture. One of the three phosphates on the end of the molecule can separate, releasing energy. The depleted ADP (adenosine diphosphate) is then recycled back to the shop to be recharged by replacing the third phosphate group.
  * The bottom illustration is the structure of the mineral apatite (pronounced "appetite"). Our bones and teeth are made of apatite. Apatite is a calcium phosphate with fluoride (F⁻), chloride (Cl⁻), or hydroxyl (OH⁻). The illustration shows fluorapatite with the fluoride ions shown in green. Fluorapatite is the least soluble of the three choices; we try to get trace amounts of fluoride ion to young children to reduce tooth decay.

Previous Page:
Three major biological roles for phosphate.

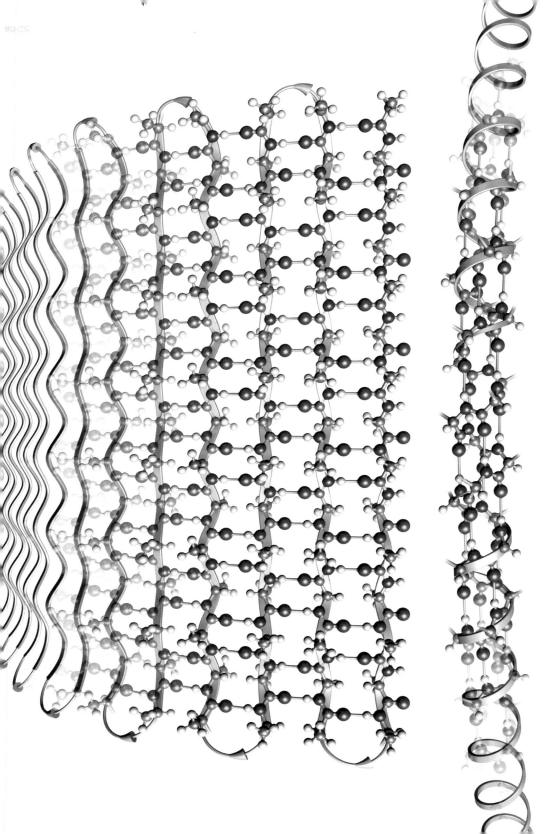

13   ALPHA HELIX AND BETA SHEET

In eight scientific papers published in 1951, Linus Pauling and his coauthors established the structures now known as the *alpha helix* and the *beta sheet*. These structures are composed of proteins: chains of amino acids. The string of DNA instructions can lead only to a linear string of amino acids. Curling that long protein string into a useful structure is a major accomplishment of living systems.

Pauling's 1951 insight showed that hydrogen bonds (as in ice, #2) could link the amino acid chain into alpha-helix rods and beta sheets. Hydrogen bonds are not as strong as covalent, ionic, or metallic bonds, but they are strong enough to hold together biological structures. Much of the specific geometry required for protein structure is supplied by alpha helixes and beta sheets.

Because alpha helixes and beta sheets make up a large percentage of most protein structures, biologists often use *ribbons* as a shorthand in illustrating protein structure. The alpha helix is shown on the left-hand side of the illustration with the atoms fading out at the top and bottom and the ribbon symbol continuing. Similarly, the ribbon depiction continues onward for the beta sheet on the right.

Beta sheets are assembled by folding the protein chain back and forth. In early Etruscan and Roman writing, alternate lines of text reversed direction, an arrangement known as *boustrophedon*, "as the ox plows." Beta sheets use boustrophedon, as shown by arrows on the top and bottom of the illustration. Beta sheets are not geometrically flat. In the illustration, the beta sheet turns around a corner to show the possibility. There are various barrel, sandwich, and propeller arrangements of beta sheets.

This particular beta sheet comes from silk made by a wild silkworm moth. Both silkworm silk and spider silk are remarkable materials – very strong for their weight. Synthetic fibers have not put the commercial mulberry silkworm out of business.

A view down the axis of an alpha helix is shown on the next page. The helix is a right-handed spiral. (Ordinary bolts and wood screws are right

Previous Page:
Hydrogen bonds stabilizing the protein structures of an alpha helix and a beta sheet.

handed.) In that end-on view, each amino acid advances through a turn of 100 degrees. Three amino acids make up 300 degrees, but that is 60 degrees short of a full 360-degree rotation. (Pauling's predecessors had been searching for a full 360-degree rotation.) In some proteins, several alpha helixes are twisted together in a ropelike composite – a coiled coil. Our fingernails, hair, and tendons are built up from alpha helixes. Boiling in water partially disassembles alpha-helix structures, which brings us gelatin, marshmallows, and smelly old-fashioned carpenter's glue.

In 1951, Linus Pauling was coming under enormous political pressure because of his opinions on nonscientific topics. The day after he sent off the alpha and beta papers for publication, the House Un-American Activities Committee issued a press release saying that Pauling helped organizations that had "complete subservience to the Communist Party of the USA, and to the Soviet Union." The dispute deepened in 1954 after the frightening Castle Bravo hydrogen bomb test in the atmosphere and because Pauling won the Nobel Prize for Chemistry.

A resolution, of a sort, happened in 1962 and 1963. The Cuban missile crisis of October 1962 caused most Americans to think of thermonuclear war as a real possibility. In 1963, the Limited Test Ban Treaty stopped nuclear testing in the atmosphere and it was announced that Pauling had won the previous year's Nobel Prize for Peace. Within a week after the prize was announced, Pauling resigned from the Cal Tech faculty. There are multiple interpretations of the circumstances for his resignation. One of Pauling's former students quizzed a number of senior Cal Tech faculty members and was told that the federal government had informed the Cal Tech administration that Pauling either had to shut up, or our beloved U.S. government would cancel every Cal Tech research grant, graduate student fellowship, and undergraduate loan, as well as the management contract for the Jet Propulsion Lab. Given the threat, Pauling decided to resign.

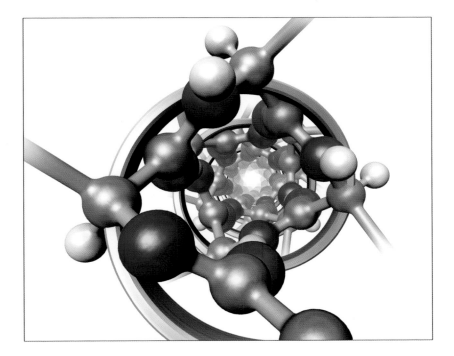

Above:
    End view of an alpha helix.

14   LYSOZYME

Enzymes are particular kinds of globular proteins that are the biological equivalents of catalysts in chemistry – something that speeds up a chemical reaction without itself being consumed. Lysozyme is a defensive weapon, made from a chain of 129 amino acids. It cuts a specific chemical bond that occurs on the outside of about half of all bacteria. This enzyme opens, lyses, the bacterial cell wall – hence the name lysozyme. It is the protein with the best determined structure, in part because it crystallizes well and it makes up about 3 percent of the weight of the egg white in chicken eggs, available in your local grocery store.

This illustration of the molecule shows that the dark, hydrophobic, amino acids cluster in the center and the light-toned hydrophilic ones are on the outside, just as Kauzmann had predicted (#11). The active site that does the actual work, the business end, is in the notch on the right-hand side. In the notch, two amino acids are poised to grab, and break, a specific weak spot in the bacterial cell wall.

In the illustration carbon atoms are represented as gray, oxygen atoms as red, nitrogen as blue, and sulfur as yellow. As in the preceding amino-acid illustration the hydrophobic atoms are dark, the hydrophils are light.

In addition to the hydrophobic-hydrophilic guidance, there are four internal bonds between sulfur atoms on pairs of the amino acid cystein. Not all of the four sulfur-sulfur (cystein) pairs are visible in the illustration; some of them are behind other atoms. Lysozymes from dozens of sources other than chicken eggs have been determined, and there are lots of mutations that change the amino acid sequence. However, none of the variants has any changes in the eight cystein locations. Apparently, any substitution or relocation of the cysteins is fatal.

Previous Page:
    A globular protein, lysozime, showing the water-loving (light-toned) amino acids on the outside.

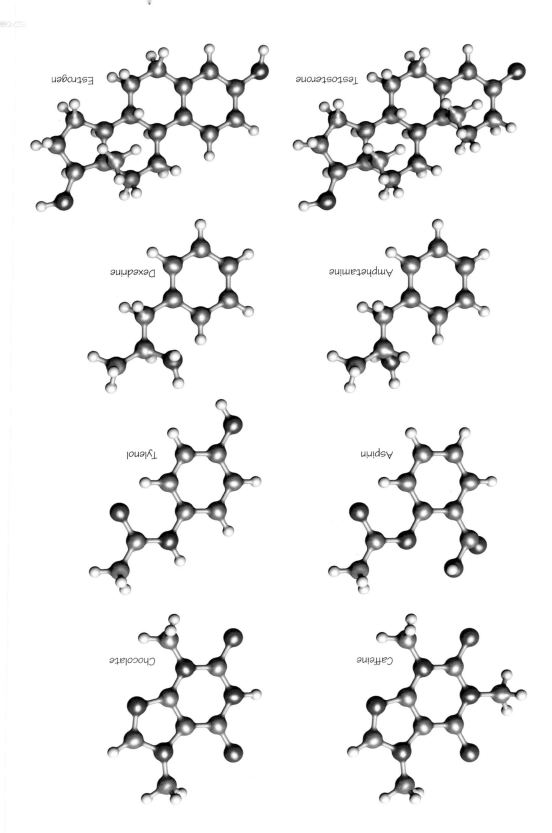

15 DRUGS

Most large biological molecules get broken up in our digestive systems. Some small molecules get through, but minor differences between closely related molecules can result in different effects. We show four pairs of molecules, using our normal color scheme with white hydrogens, red oxygens, blue nitrogens, and black carbons.

At the top, the caffeine in coffee differs from the major flavor in chocolate only by having one methyl (CH_3) group added on the left-hand side.

Tylenol and aspirin are used for roughly the same purposes, and the molecules are similar but not identical. In detail, aspirin and Tylenol have different effects. In 1917, the U.S. Supreme Court ruled that aspirin was a generic drug. (The patent owner, Bayer, was a German company and World War I was raging.) In contrast, Tylenol is a trade name owned by the McNeil division of Johnson & Johnson.

Amphetamine and Dexedrine are used as "wake-up" drugs, but Dexadrine has a reputation for being less addictive. The molecules are identical, except that they are mirror images of one another. The nitrogen shown on the upper side of the molecule faces toward the viewer in Dexedrine and points away from the viewer in amphetamine. Both are available as prescription drugs, but a slightly different molecule (methamphetamine) is widely produced as an illegal drug.

At the bottom of the page, the hormones estrogen and testosterone differ only in having an extra methyl group in testosterone and one added hydrogen in estrogen. Testosterone makes boys and estrogen makes girls. *Vive la différence!*

Previous Page:
Structural similarities between pairs of biologically active molecules.

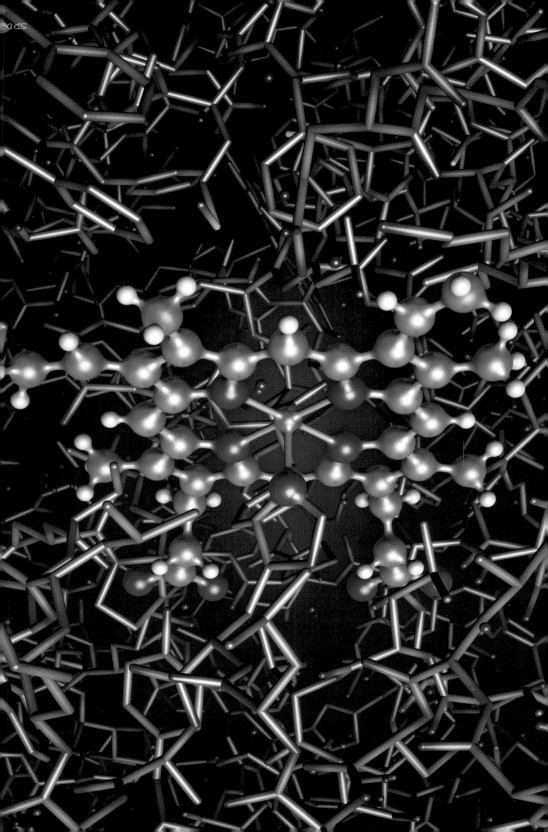

16   HEMOGLOBIN

Hemoglobin is the red in our red blood cells. It's a carrier for oxygen. Although a red blood cell takes only five seconds to go through the lung, it gets fully loaded with oxygen. Oxygen is not the only gas that attaches to hemoglobin; carbon monoxide is hazardous because it attaches strongly to hemoglobin. Some red meats and high-grade tuna are treated with carbon monoxide to enhance the red color. Although it sounds bad, the carbon-monoxide-treated meat is not a health hazard to people.

Hemoglobin is a compound word. The *heme* part is a molecular ring with an iron atom at the center. The oxygen molecule being transported sticks to the iron. In humans, four heme rings are distributed inside a globular protein; that's the *globin* part of the name. The illustration to the left shows the heme ring as a ball-and-stick model with blue nitrogen atoms and a gold-colored iron atom at the center. The immediate surroundings inside the globular protein are shown as a stick model.

Hemoglobin is not the only possible oxygen carrier. A copper-containing molecule, with a structure completely different from the heme ring, occurs in *Limulus*, the horseshoe crab. Small amounts of the crab blood, blue when oxygenated, can be donated without killing the crab. The blood is important for certain medical tests and it retails for $15,000 a quart. Aristocrats share their "blueblood" designation with some bottom-feeders.

Among humans, there are some small variations in the protein globin part of hemoglobin. Sickle-cell anemia, which confers a partial resistance to malaria, is caused by a single substitution in the protein. Sickle-cell anemia was the first human disease to be diagnosed right down to the molecular level. Guess who discovered it: Linus Pauling, in 1949, before Watson and Crick discovered the importance of DNA.

When the hemoglobin reaches the delivery sites, it hands off the oxygen to mioglobin. By now, you decipher the *globin* to be a globular protein; the *mio* part is an even hungrier iron-containing ring that pulls oxygen away from the hemoglobin. Some serious fine-tuning is apparent: Hemoglobin has to bond to oxygen strongly enough to absorb oxygen in the lungs, and weak enough to

Previous Page:
    A portion of the globular protein hemoglobin, with the heme
    ring highlighted.

surrender it to mioglobin in the muscles. The isolated heme ring illustrated on the next page shows four blue nitrogen atoms in a ring around the iron, and one additional nitrogen atom on one side of the ring. The fifth nitrogen plays an important role in fine-tuning the oxygen affinity. We have shown the heme ring in its flat "working" configuration for carrying oxygen. After delivering up the oxygen, the heme is slightly puckered.

Because iron plays a major role in hemoglobin and mioglobin, iron is an essential nutrient for most organisms. In seawater, organic growth is sometimes limited by the availability of iron. In a cubic mile of ocean, there is often more iron inside the fish than in all the surrounding seawater.

Multicelled organisms first appear, rather abruptly, in the geologic record about 540 million years ago. (To a geologist, "abruptly" can mean 50 million years.) Except for vertebrates, all of our modern phyla – the basic body plans – appeared during what is called the *Cambrian explosion*. Schemes for transporting oxygen and using it to metabolize food had to appear at that time. It isn't clear whether the oxygen schemes were invented independently many times or whether it was invented only a few times and the secret was passed around by viruses. It couldn't have been invented only once; remember the horseshoe crab.

When I was running California fall-break field trips for Princeton freshmen, one of my favorite stops was at the appropriately named Fossil Gulch in the White Mountains. (Sometimes the students weren't as thrilled as I was.) The sedimentary rocks were deposited close to the beginning of the Cambrian explosion. There were worm tracks in the rock, meaning that something multicelled was working. The only body fossils were Archaeocyathids, probably primitive sponges, about an inch across. I called it the "ka" in the "ka-boom" of the Cambrian explosion.

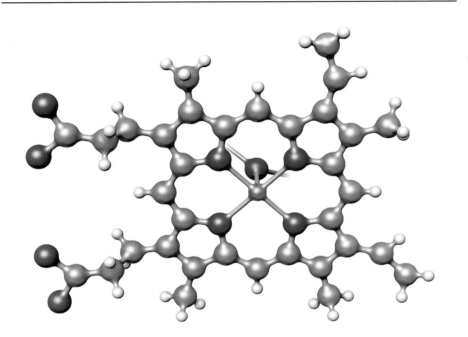

Above:
    Enlarged view of the isolated heme ring from hemoglobin.

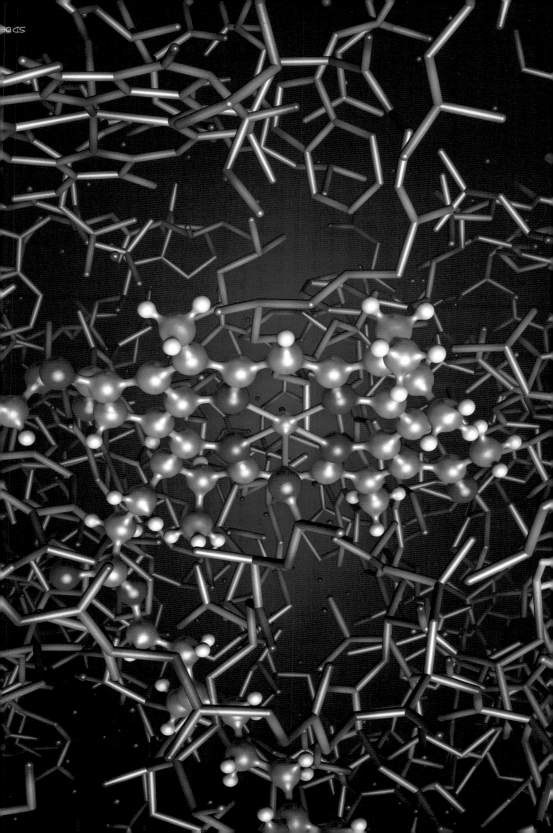

17   CHLOROPHYLL

Blood circulation has been around for "only" 540 million years, but photosynthesis has been active for about 3 billion years. Photosynthesis uses sunlight, carbon dioxide, and water to produce high-energy carbohydrates (sugars and starches). The very oldest undeformed rocks show signs of photosynthetic activity. My generation of geologists was taught that there were essentially no signs of life more than 540 million years old. When I was teaching at the University of Minnesota during the 1960s, there were some handsome rock specimens around showing photosynthetic colonies 2 billion years old; that was not supposed to happen. We finally brought in a pickup-truck load of the rock so that all academic visitors to the geology department could take home a rule-breaker.

The most primitive photosynthetic organisms are grouped as *cyanobacteria*. (Formerly, they were called *blue-green algae*.) Cyanobacteria include the dark stuff that grows in wet corners of a shower stall. In today's tropics and subtropics, flats that are inundated with seawater only at the highest tides are often covered with mats of cyanobacteria. I feel that I am walking on sacred ground; these mats are the oldest lifestyle on the planet. It's a supremely tough environment. Seawater from the previous highest tide evaporates until salt crystals grow. Then comes an afternoon thundershower and in a few minutes the bacteria find themselves in pure water. The osmotic pressure change is roughly a ton per square inch (9 megapascals) (see p. 131). My opinion: No other life forms can cope with the abrupt environmental changes.

The illustration shows the chlorophyll molecule, embedded in its protein housing. A ring of four nitrogen atoms (blue) surrounds a magnesium ion (green). A chlorophyll molecule is essentially an antenna for grabbing red and blue light; the unused reflected green light accounts for the green color of leaves. From the chlorophyll, a succession of complicated biochemical steps eventually produces carbohydrates. (When I try to read about the steps, I get the giggles because a key enzyme is named RuBisCO; sounds like something from an advertising agency.)

Magnesium, used in chlorophyll, is an essential plant nutrient. There are magnesium-deficient soils, especially sandy soils. I had a short and unsuccessful career looking for magnesium-deficient soils. Along the margins of

Previous Page:
The heme ring of chlorophyll, highlighted within its protein setting.

the Snake River Plain in Idaho are some volcanic rocks with exceptionally low magnesium content. A soil scientist went with me and we scoured one of the localities looking for distressed vegetation. Nuttin', zip, nada; the plants looked healthy. My one-day career in soil chemistry ended in total failure.

Chlorophyll is not the only natural light-gathering molecule. The most spectacular examples give the name to the purple bacteria; they absorb green. One fascinating speculation was published: The purple molecule may have originated first and chlorophyll could have filled in to use the light the purple bacterial rejected. Green chlorophyll might have been the underdog that emerged the winner.

If you have flown into San Francisco airport, you may have noticed the southeast end of San Francisco Bay divided into large seawater evaporation ponds that recover salt. Ponds are colored lingerie pink and lipstick red by purple bacteria. I'm going to miss them; the ponds are being decommissioned.

The illustration on the next page shows the chlorophyll molecule stripped of its long hydrocarbon tail and without some of the methyl groups that hung around the edge. These "stripped down" chlorophyll molecules are found in almost all crude oils, derived mostly from cyanobacteria. The molecules are known collectively as *porphyrins*. The magnesium at the center is typically replaced by copper, nickel, or vanadium. This illustration shows vanadium at the center, with oxygen atoms above and below. The oxygen-vanadium-oxygen complex is known as *vanadyl* and it is particularly abundant in heavy crude oil from Venezuela. Turbine engines, such as aircraft jet engines, have a problem with vanadium, but additives to the fuel have been developed that avoid the expense of removing the vanadium at the oil refinery.

Next Page, Top:
  Heme ring, stripped out of chlorophyll, as found in crude oil, with the central iron atom replaced by a vanadyl (VO_2) group.

Next Page, Bottom:
  Salt ponds in San Francisco Bay were colored pink and red by "purple" bacteria.

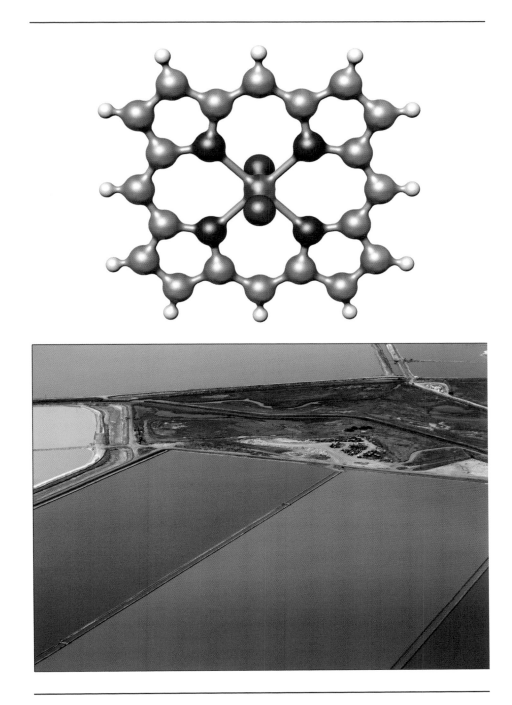

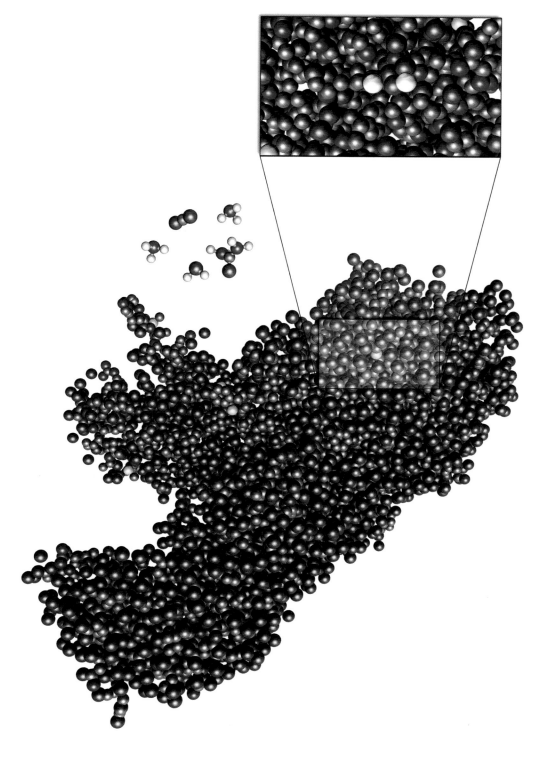

18 UREASE

Before 1828, an absolute barrier was thought to exist between the chemicals produced by living organisms and "inorganic" chemicals produced in the lab. Vitalism was thought to be unique to living systems. That barrier was broken by Friedrich Wohler, who produced urea in his lab from inorganic chemicals. Urea was named, obviously, from urine. Most of the protein nitrogen in the adult diet winds up as urea in the urine.

A few billion years before 1828, bacteria developed a method for running the urea synthesis backward: converting urea into carbon dioxide and nitrogen-containing ammonia. The bacterial reaction was speeded up by a globular protein called *urease,* used as a catalyst. Enzymes can be sorted into two great divisions: those that catalyze acid-base reactions and those that catalyze oxidation-reduction reactions. An old saying goes, "Let's you and him fight; I'll hold your coat." In acid-base reactions the catalyst holds one or more hydrogen ions while the reactants and products fight it out. The amino acids that make up proteins have a wide variety of affinity for hydrogen ions. In contrast, proteins have essentially zero capacity for holding and releasing electrons, which are the root cause of oxidation and reduction. Usually, oxidation-reduction enzymes incorporate atoms (not found in amino acids) that are capable of reducing or increasing their oxidation states. These atoms hold the electrons while the oxidation-reduction reactants and products fight it out.

The usual choice in oxidation-reduction catalysts is a pair of two identical atoms from the group known as the transition metals: chromium, manganese, cobalt, nickel, copper, zinc, molybdenum, and cadmium. On the label of a bottle of multipurpose "vitamins and minerals" there is often a list of unfamiliar metals, most of them for use in oxidation-reduction enzymes. Even though all organisms need these metals at the parts-per-million level, there are examples of trace-metal deficiencies. One that caught my attention was a "before" picture of a scraggy pasture in New Zealand. The "after" picture showed abundant plant growth after less than a dollar's worth of molybdenum was applied to each acre.

Previous Page:
  *The urease enzyme is shown with an enlarged box uncovering the two
  nickel atoms. A molecule of urea and a water molecule are arriving and
  a carbon dioxide and two ammonia molecules are leaving.*

In urease, the metal of choice is a pair of nickel atoms. The two nickel atoms are close together but obscured from view by amino acids in the globular protein. In the enlarged section of the accompanying illustration, we have peeled off some of the overlying amino acids to show the two nickels. Those of us old enough to have raised children back in the cloth-diaper era remember the sharp odor of ammonia when we lifted the diaper pail lid. One possibility would be to keep even tiny sources of nickel out of the diaper pail. It's a losing battle; the stainless steel in the washing machine is a nickel alloy.

Urea has an industrial role. One use is for urea-formaldehyde plastics. About 30 years ago, one group modified the urinals in Yankee Stadium to flush with very little water and the urea was recovered to make plastics. (There was no record about whether they subsidized the price of beer.) Today, urea has a major use as a nitrogen-containing fertilizer. The urea is made inorganically from ammonia and carbon dioxide. The ammonia is produced by reacting atmospheric nitrogen with hydrogen. The hydrogen normally is generated from natural gas, although some factories in China are using the steam-coal reaction as the hydrogen source.

Urea can be transported in bags as a dry component of fertilizers. Urea is a favorite nitrogen source for plants because the bacteria take some time in converting the urea into the form of nitrogen that plants absorb; it's a one-molecule timed-release fertilizer. In contrast, nitrogen supplied as nitrate can be washed out in the very next rain.

Above:

    Urea, manufactured from natural gas or from coal, is a preferred carrier of the nitrogen component in fertilizer.

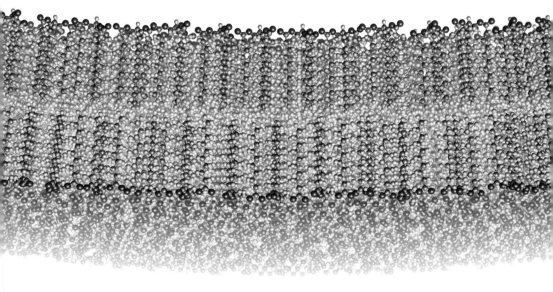

19 LIPID MEMBRANE

The outer cell walls of simple single-cell organisms typically consist of a double layer. The building blocks of the layer are called *lipids*, which consist of a water-loving glycerine component on one end and two or three water-hating hydrocarbon chains attached. All biologically produced oils, fats, and waxes are lipids. You have lipids coming out your ears. The illustration shows a segment from a bacterial cell wall with the glycerine endings both on the outside (above) and on the inside (below). Longer hydrocarbon chains make the cell wall an even tighter seal against chemical transport.

Glycerine, more properly called glycerol, is a triple alcohol with three hydroxide (OH⁻) groups. In many lipids, one of the three hydroxide sites is occupied by a phosphate group and the other two by hydrocarbon chains. However, in simple marine organisms the ratio of carbon to phosphate is typically 106 to 1, which is a strong hint that most of the cell-wall lipids do not contain phospholipids. This result, which was discovered in 1934, caused us to choose a lipid with three simple hydrocarbon chains for the illustration.

Cell-wall lipids play a major role in generating crude oil in the ground. I explained the process on pages 14–21 of *Hubbert's Peak* (Princeton University Press, 2001) and I don't need to repeat it here. A few skeptics insist that natural gas, and even oil, is being generated by inorganic processes deep in the earth. My negative comment on inorganic oil is on pages 62–63 of *Beyond Oil* (Hill and Wang, 2006) and I have nothing further to add at this time.

An intriguing hint about the origins of lipid membranes comes from the tendency for lipids to organize themselves spontaneously into lipid bilayers. Studies first reported in 1977 and 1978 showed that, above a certain minimum concentration, lipids would join together in curved surfaces, including closed spheres. This by no means proves that lipid cell walls originated spontaneously, but at least it suggests that cell walls do not necessarily require a pre-existing cell to build them.

Previous Page:
  Atomic view of a lipid membrane, with an enlarged view of the triglyceride building block shown above.

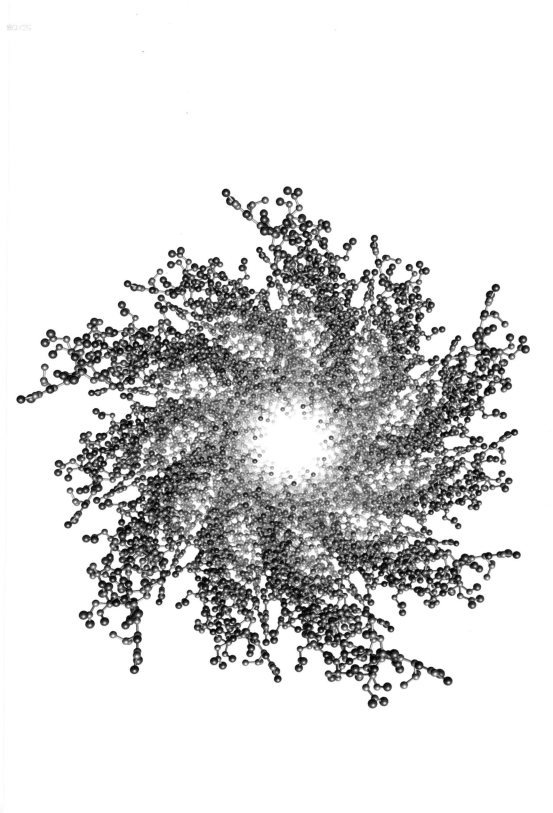

20 ROD VIRUS

A virus, by itself, has no metabolism and cannot reproduce. They function only by invading intact living cells – from bacteria to humans – and by commandeering the cell's machinery and materials to churn out copies of the original virus. Viruses are a kind of mobile parasite. A chicken is just an egg's way of making another egg. To a common-cold rhinovirus, I'm just a way of creating another rhinovirus.

In that sense, viruses are a huge success. There are lots of individual virus particles floating around, and there exists a gigantic diversity of different kinds of virus. Recent analyses of seawater suggest that a big fraction of all viruses remain to be described. Animals, plants, fungi, and bacteria have their own specialized viral parasites.

The illustration is an end-on view of a fibrous virus. To be honest, we chose it because it made an attractive illustration. A side view on the next page shows that the virus resembles a sheaf of wheat. This particular virus was grown in cells of *Pseudomonas aeruginosa*, a bacterium that is a frequent cause of hot-tub rash.

The basic structure of a virus consists of an outer *capsid*, a coat of repeated protein units. Inside is the genetic material. Viruses use all manner of genetic materials, double-stranded or single-stranded, DNA or RNA, and several replication schemes. The virus strategy requires locating the surface of a susceptible cell, injecting the viral genetic material into the target cell, producing hundreds of copies of the genetic material, generating additional capsid proteins, and escaping from the cell. Sometimes "escape" means killing the cell.

Mutation rates (changes in the genetic material) are very rapid in viruses. We higher animals have elaborate editing mechanisms to correct errors. Because there are so many individual virus particles, sometimes viruses test the joint effect of two simultaneous mutations. People, as biologically conservative entities, typically can test only one mutation at a time.

The genetic material in the largest viruses is smaller than the smallest bacterial genome. The largest viruses have about 100 genes, where one gene essentially means one protein and most of the proteins are catalytic enzymes.

Previous Page:
View down the length of a fibrous rod virus.

The smallest bacterial genome contains 182 genes, essentially 182 enzymes. In the first edition of his 1985 book *The Origins of Life*, the physicist Freeman Dyson estimated that the smallest viable living cell would require at least 100 enzymes. Good shooting; that estimate falls right in the gap between the largest nonliving viruses and the simplest bacterial cell. Smaller viruses exist; the smallest has only four genes.

Only a poorly adapted parasite would promptly kill its host. The terrifying Ebola and Marburg viruses described by Richard Preston in *The Hot Zone* kill people so quickly that the victim has few opportunities to spread the virus around. In contrast, the HIV-AIDS virus kills slowly enough to allow the virus to spread. Yet more efficient is the common cold virus, which spreads readily and almost never kills the host.

The long rod-shaped virus shown on the next page is the tobacco mosaic virus, so named because it causes a mosaic mottling of tobacco plant leaves. It was the first virus to be identified and studied. Actually, the geometry is a helix, a long thin spiral. The lumps on the outside of the virus are capsids. One of the particles shows the internal genetic material that can be injected into a cellular victim.

Above:
    Side view of the fibrous rod virus shown on the previous page.

Above:
    Tobacco-mosaic virus cylinders.

21   ICOSAHEDRA VIRUS

Icosahedral viruses include thousands of different types. The icosahedron was known to the ancient Greeks: a solid bounded by twenty equilateral triangles. The icosahedron was used by Buckminster Fuller in his geodesic domes, it occurs as carbon buckyballs, it is the geometry of older soccer balls. Early nuclear weapons made use of the symmetry.

In the illustration, Stephen has included white lines to emphasize the symmetry. (Real viruses do not have an aluminum tube framework.) The colors serve to identify individual capsid proteins on the surface; the genetic material is protected inside. The structure of one of the capsid units is shown at the atomic level. When all of the capsids are shown at the atomic level, the virus looks like an enormous unstructured fuzzball.

The particular virus in the illustration causes polio, a once-dreaded disease. The most famous polio victim was President Franklin Roosevelt, who spent the last 24 years of his life in a wheelchair. As I can vividly remember, the 1952 introduction by Jonas Salk of an injected polio vaccine and the 1958 discovery by Albert Sabin of an oral vaccine were medical victories of the first magnitude. Today, polio has been eradicated worldwide, except in portions of India, Pakistan, Afghanistan, and Nigeria, and work continues in those four countries.

There are human diseases that are caused by bacteria and fungi, but even a partial list of viral diseases is impressive: rabies, smallpox, yellow fever, polio, flu, chickenpox, hepatitis, measles, mumps, herpes, HIV-AIDS, and the common cold. Development of the polio vaccine took about ten years, funded by the March of Dimes campaign. HIV has turned out to be a much more stubborn opponent; 25 years have elapsed without the development of a preventative vaccine.

It comes as no surprise that unwelcome bits of computer code are called viruses. A proper classification of computer diseases recognizes worms and Trojan horses as well as viruses.

Previous Page:
The polio virus, shown with the protein-cover units (capsids)
distinguished by different colors. One unit is detailed at the atomic level.

22
UNIT CELL DISCOVERY

This is one of those legends that are probably true, but even if it isn't factual it ought to be true. Around the year 1800, Rene Just Haüy owned an extensive assortment of calcite crystals (calcium carbonate, $CaCO_3$). Haüy accidentally dropped a friend's calcite crystal and it shattered into small twinkly pieces. He recognized that the shattered pieces had the same shape as a particular calcite crystal that he had at home. Haüy went home and smashed his entire calcite collection; all of them broke into the same-shaped pieces as his friend's crystal.

Haüy inferred that the shattered pieces were the expression of some kind of internal building block. The diversity of external faces on calcite crystals could result simply from stacking rules: "Down two, over one." On the upper left, we repeat Haüy's famous drawing, published in 1801, which showed the stacking rule for the calcite crystal form known as dogtooth spar. (In polite society, the shape is called a *scalenohedron*.)

After diffraction of x-rays by crystals was discovered in 1912, calcite was one of the early crystal structures to be determined at the atomic level. In the larger illustration, we revise Haüy's diagram by showing the atoms in place. (Oxygen is shown in red, calcium in white; the black carbons are usually hidden behind the surrounding oxygen.) The x-ray diffraction pattern of a calcite crystal is shown on the following page.

The concept of a unit cell, a building block, turned out to be universally true for crystals. External crystal faces, expressed as stacking rules for unit cells, are given numbers known as *Miller indices* for the British mineralogist William Miller. However, rumor has it that Miller popularized a method that was originally devised by William Whewell (pronounced "Hewell"). Haüy's use of the cleavage rhomb of calcite differs slightly from the preferred modern mathematical system for assigning unit cells, but his building block interpretation is intact.

The three-number Miller indices describe a crystal face by giving the count of unit cells along three axes: north-south, east-west, and up-down. An Egyptian pyramid has Miller indices 111, dogtooth spar is 214. That works

Previous Page
(Upper left) Haüy's 1801 drawing of the calcite building blocks and (center) the same structure at the atomic level.

for thousands of different minerals and tens of thousands of lab-created crystals. However, one mineral stands out as an exception. The mineral is calaverite, named for the California county made famous by Mark Twain's frog. Calaverite is gold telluride, $AuTe_2$. The chemistry is simple but the structure is bizarre. Crystal faces on calaverite require four Miller indices: north-south, east-west, up-down, and don't ask. Calaverite is projected down to us from a four-dimensional space into our familiar three spatial dimensions. I can't visualize it. But if you want to start a new fad in crystals worn on a neck chain, the major source of calaverite today is at Cripple Creek, Colorado.

For the first thirty years after x-ray diffraction was discovered, the equipment was quite cumbersome, it was hand-made, and it had considerable safety problems. In 1945, the Dutch company Phillips introduced a manufactured x-ray diffraction device. Although it weighed several hundred pounds and took up about 25 square feet of floor space, it became a standard tool in geology and in chemistry departments. Gradually, a library was gathered recording the x-ray patterns from 130,000 natural and synthetic crystal structures. It's now called the Powder Diffraction File. Positive identification of a mineral usually took one hour.

I learned the strength of x-ray diffraction the hard way. As a graduate student in 1956, I was using a Phillips diffraction unit to identify zeolite specimens from Nevada that looked like blackboard chalk. Because a library of known x-ray patterns was not then available, I received permission to chip off small corners from the zeolite crystals in the Princeton museum to use as standards. Enormous confusion ensued. Gradually, I realized that half of those beautiful museum specimens were identified incorrectly; identified incorrectly by professors who wrote mineralogy textbooks. About that time, I heard that Harvard was running x-ray patterns on every specimen in their huge collection; they had similar identity problems. X-ray wins, mineralogy professors lose.

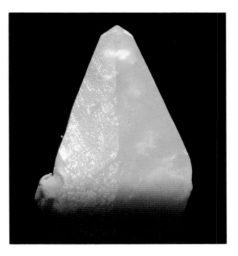

Right:
   A photograph of
   the "dogtooth spar"
   crystal form of calcite.

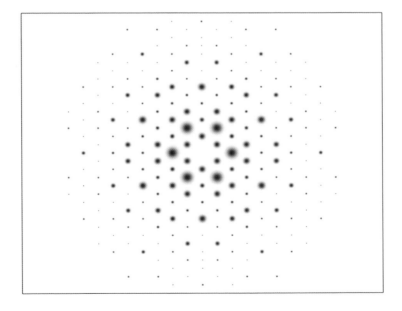

Above:
   The x-ray diffraction pattern of a calcite crystal.

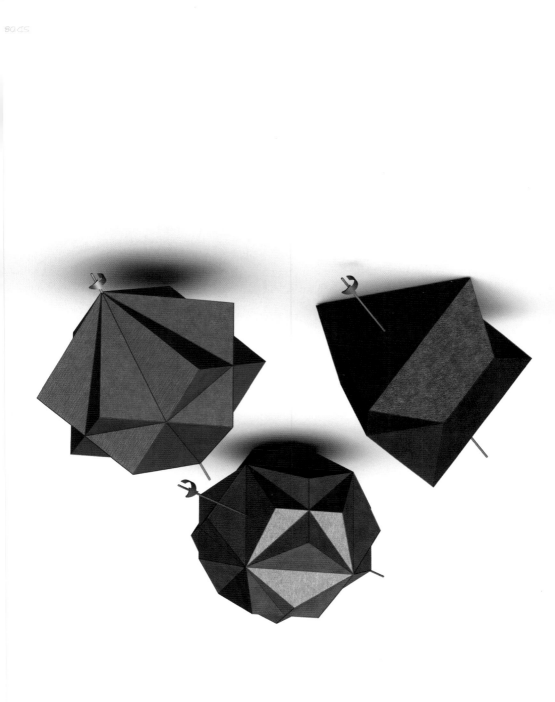

23   TWINNED CRYSTALS

A crystal twin adds a symmetry element that is not present in either half of the twin. Most twins can be described by a rotation of 180 degrees about an axis. (Although a mirror plane might seem to be an exception, a mirror plane is a rotation of 180 degrees plus a center of symmetry.) The 180-degree axis of rotation is shown in the illustration for each of the three twinned crystals. Twins are described in several categories. A contact twin has an obvious planar boundary between the twinned segments, as in the spinel twin. In contrast, the fluorite and pyrite twin segments interpenetrate; they are called penetration twins. Twins are also divided into growth twins and deformational twins.

All three of the twins shown are from the cubic system, although twins can – and do – occur in crystals of any symmetry.

* On the lower right is a penetration twin typical of the mineral *fluorite* (CaF_2), but it also occurs in other cubic crystals. A cube is rotated about its body diagonal. The name fluorite comes from the Latin root *fluo*, to flow. The Romans used it to make their glass and smelting fluxes more fluid. In turn, fluorite gave its name to the chemical element fluorine and the fluoride ion. When exposed to ultraviolet light, most specimens of fluorite give off visible light. Although fluorite is not the only substance to do so, it did lend its name to the property – fluorescence (and hence, fluorescent lighting, which does not use fluorite at all). Most recently, fluorite has turned up in high-quality telescopes and in photographic lenses. The words are often spelled incorrectly. Flour and flout are something else.

* On the lower left is an octahedron (originally with eight triangular faces), with the same rotation axis as the fluorite twin. The front half of the crystal has been rotated, while the rear half stayed still. This contact twin pattern is called a *spinel twin*. When the same twin occurs in diamond, it is called a *macle* and it presents special problems for diamond cutters. The original spinel mineral contains magnesium, aluminum, and

Previous Page:
Three different styles of twinning in cubic crystals.

oxygen ($MgAl_2O_4$). However, a long list of crystals – natural and synthetic – share the spinel crystal structure. Virtually all the world's chromium comes from chromite, which has the spinel structure. The world's most famous spinel was originally misnamed the "Black Prince Ruby." Actually, it isn't a ruby at all; the label in the British crown jewels now says "Black Prince Ruby Spinel."

❋ At the top is a penetration twinned crystal of pyrite (iron sulfide). For obvious reasons, it is called the "iron cross" of pyrite. (The iron cross medal was initiated in the Prussian army, long before World War II.) Another of my dirty exam questions asks about the electric charge (called *valence*) on iron and sulfur in pyrite (FeS_2). Pyrite tends to form in environments that lack oxygen, so the iron would be expected to be Fe^{++} and the sulfur S^{--}. The logical compound would be FeS – the mineral troilite, which occurs frequently in meteorites and almost never on Earth. The answer to the dirty question is that one of the two sulfur atoms in pyrite is the expected S^{--}, and the other sulfur has no electric charge S^0. The electrons that carry the electric charge can switch from one sulfur to the other, which gives pyrite its metallic look and nickname: fool's gold.

An example of growth twinning comes from evaporating a drop of seawater on a microscope slide. Fishtail-shaped twins of gypsum ($CaSO_4·2 H_2O$) can be seen through the microscope. A do-it-yourself deformation twin of calcite is shown in the following section.

Sodium-calcium feldspars typically show repeated twinning, back and forth, that divide the feldspar into parallel sheets, called *lamellae*. The phenomenon achieved 15 minutes of fame when the astronauts on the Apollo 15 mission confirmed that the lunar highlands were dominated with calcium feldspar by saying, "I can see the twin lamellae from here."

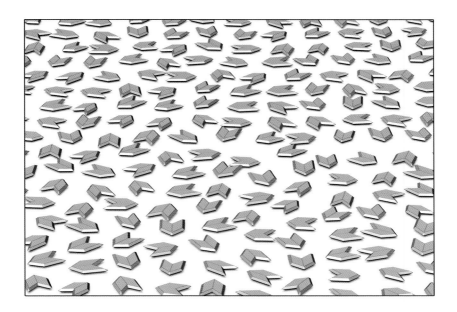

Above:

"Fishtail" twinned gypsum crystals from evaporated seawater.

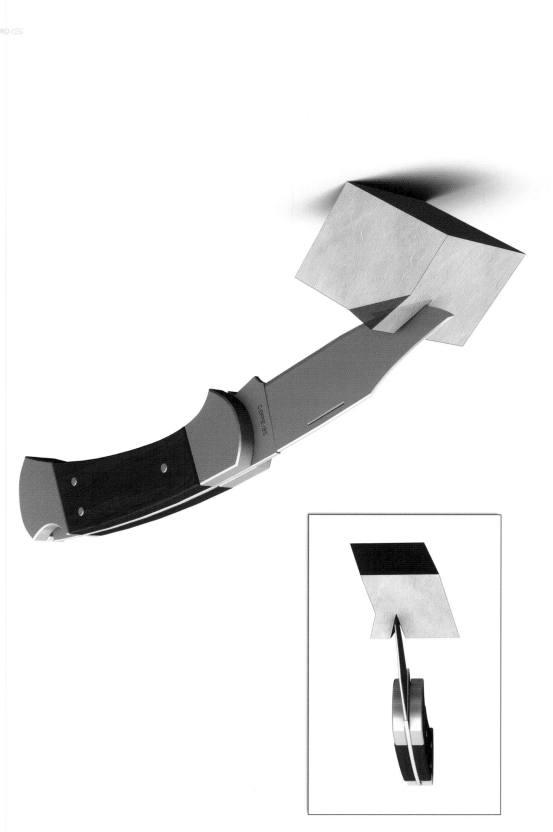

We felt that a twinning experiment on a natural crystal, at room temperature, would remove a little of the mystery from twinning. It seems less exotic if you can do it on a tabletop, powered by your fingers.

The experiment begins with a cleavage fragment from a calcite crystal. Calcite is calcium carbonate. (A different structure, also made of calcium carbonate, is called *aragonite*, named for the province that brought you Catherine of Aragon.) Calcite is found in antacid pills, ornamental stone (including marble), and kitchen scouring powder. About half of all seashells are calcite; most of the rest are aragonite.

Crystalline calcite is available from several mineral dealers. Clear, optical grade, calcite is beautiful but expensive; ordinary cloudy calcite will work equally well. Use a hammer, or a chisel and a hammer, to break out a cleavage rhomb at least a half inch in each direction. (The size doesn't matter for the science, but if you are holding it with your fingers you don't want to get cut during the next step.)

If you hold the cleavage rhomb with one edge down against the table top, the rhomb will look tilted over to one side. (See the end-on view in the small box on the upper left.) We use a knife blade to flip over a small part of the upper corner to the opposite tilt. "Small" is important. You can flip over a millimeter-size piece of the upper corner.

What you have done is to create a contact twin. The same thing happens when rocks are deformed by natural forces. What happens at the nano scale is shown in the following section.

Previous Page:
   A calcite twinning experiment.

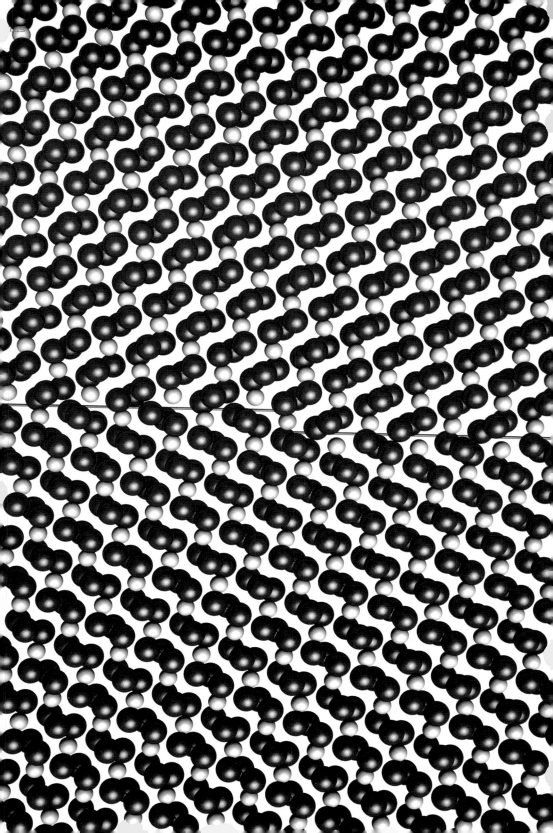

25 CALCITE TWIN PLANE

Twinning a calcite crystal with your fingers is possible because you do not have to flip over all the atoms at once; you can do it one atom at a time. This is a cross section through a twinned calcite crystal. The twin plane is marked by a thin black line halfway down the page.

Below the twin plane is the original calcite crystal, with red oxygen ions and white calcium ions. The groups of three oxygen ions, tightly bound to a small carbon ion at the center, are called *carbonate* groups. Above the twin plane, the oxygen ions are flipped into a different orientation, with the carbon ion at the center of each carbonate group staying still. The structure is still calcite.

At the center of the illustration, there is an offset in the twin plane. This is one type of *dislocation*, a local break in the crystal order. As the dislocation moves – from right to left – the twinned volume above the twin plane increases, at the expense of the original calcite below. Not only can the dislocation move one row at a time, it can move into the page, flipping over only one carbonate group at a time. This is why you can twin it with your fingers.

The reality of these dislocations was demonstrated during the 1960s at the Shell research lab in Houston. Dissolving some of the calcite in dilute acid selectively removes calcium and carbonate from around the dislocation. Microscopic pyramid-shaped pits form on the calcite surface at each dislocation. The key step was to deform the crystal slightly and then treat it a second time in dilute acid. Where a dislocation had moved away, an original pit opened into a flat-bottomed pit. Where the dislocations wound up, new sharp-bottomed pits appeared.

Previous Page:
   A slice of the atomic structure of twinned calcite.

26   DOLOMITE TWIN PLANE

Warning: The facing illustration shows what does not happen. It shows dolomite in the same view as the preceding calcite example, but the twinning is impossible.

Dolomite, as shown below the impossible twin plane, is similar to calcite but with alternating layers of calcium (white) and magnesium (green) between the carbonate layers (red). Above the suggested twin plane, flipping the carbonate groups over places calcium and magnesium atoms alternating within the same layer. Above the purported twin plane, the crystal is no longer dolomite (see p. 131). The twinning of calcite, as shown in the previous section, is forbidden in dolomite. An enlarged view of the impossibly twinned result is shown on the next page, tilted to make the layers horizontal.

Dolomite has long been a puzzle. In rocks of some ancient periods, like the Devonian, dolomite is more common than limestone (made of calcite). However, only small amounts of dolomite seem to be forming today. At one time, it was said that the two great human ambitions were to build a perpetual motion machine and to make dolomite at room temperature and pressure. That ambition was fulfilled in 1995 when Judith McKenzie discovered bacteria that precipitate dolomite.

Here is my suggested clue to the paucity of dolomite in the last few million years, called the Pleistocene Epoch. Dolomite is an important oil reservoir rock, but a statistical study showed little or no oil production from dolomite of Pennsylvanian age (299 to 318 million years ago). I quizzed other geologists, we all knew about little dabs of Pennsylvanian dolomite here and there but they seemed to be quite small. Aha! What does the Pennsylvanian have in common with the Pleistocene? They are the two biggest ice ages in the last 500 million years.

The possible connection between major glaciations and dolomite scarcity comes from the unique role of magnesium as the "swing vote" in ocean chemistry. Calcium winds up mostly as a carbonate (limestone). Potassium is favored in clays. Sodium winds up in evaporated salt beds. Magnesium swings all three ways: in clays, in evaporites, and as dolomite (a carbonate).

Previous Page:
    A slice of the forbidden twin structure of dolomite.

Small annual flows of carbon dioxide – which comes largely from volcanic action – would be expressed as cold climate (glaciations) and as a halt in the deposition of dolomite.

Some natural outcrops show highly fractured layers of dolomite alternating with unfractured limestone. The interpretation is that calcite could twin and deform whereas the dolomite was broken into small blocks. Dolomite is a favored reservoir rock for oil and natural gas, but it is not clear whether fracturing or something else favors fluid flow through dolomite. As a celebration of the importance of dolomite, a double-page view of the dolomite structure appears on the following two pages.

Dolomite has appeared on the architectural scene. Concrete made using broken dolomite fragments as the aggregate develops unusually high strength. For a number of years, the Petronas Towers in Kuala Lumpur, Malaysia, were the tallest buildings in the world. They are made of reinforced dolomite-aggregate concrete. I prefer to think of them as the untwinned Towers.

Next Page, Top:
  Enlarged and rotated view of the forbidden twinned dolomite structure.

Next Page, Bottom:
  Petronas Towers in Kuala Lumpur.

Following Pages:
  An infinite view of the dolomite crystal lattice structure.

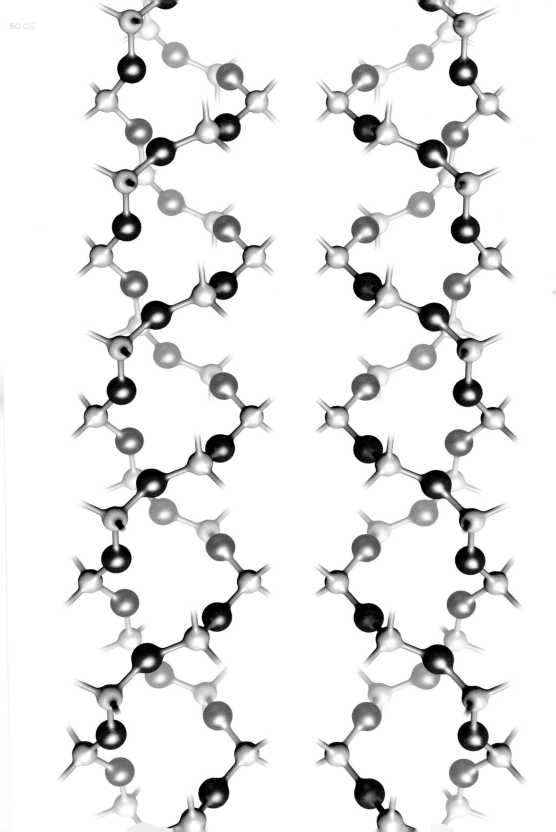

27 QUARTZ

Quartz gets around, from beach sand to wristwatches. From it we derive the silicon for computer chips and the silicone for plastics. Although quartz is silicon dioxide (SiO_2), there are many different ways of arranging silicon dioxide into solid crystalline structures. Quartz is one of the least-symmetrical structures allowed in the hexagonal system, but it's a winner. It is the most stable form of silicon dioxide at room temperature and pressure. (If the name *quartz* looks a little odd, it's because it's a medieval miner's word.)

The crystal structure of quartz lacks a center of symmetry. Some crystals like this come in left-handed and right-handed versions. We have illustrated the spiral motif from the quartz structure, with the left-handed version on the left and right-handed on the right. Left-handed and right-handed quartz are equally abundant in nature. Synthetically grown quartz crystals are right-handed, although you can buy left-handed quartz on special order.

Owing to the asymmetry in quartz, there is a correlation between mechanical stress and electrical charge: squeezing a quartz crystal generates an electrical voltage and applying a voltage causes a small change of the crystal shape. Quartz sneaked into your wristwatch because of this property. The heartbeat of a quartz crystal clock is a tiny tuning fork cut from a quartz crystal. The original quartz crystal clock in the Smithsonian fills three display cabinets; today, you can install a quartz crystal clock inside a ballpoint pen.

Quartz sand from river-deposited sediment is used in enormous ton-nages as an ingredient in concrete. Sand that has been repeatedly eroded and redeposited can contain 99 percent quartz. Clean quartz sand, especially sand low in iron content, is a major component for making glass. Silicon, separated from quartz, is used to make silicones – which are synthetic oils, rubbers, and plastics. And pure silicon is of course the favored material for electronic semiconductors. Silicon Valley was not named for the local rocks. When the electronic use of silicon was first growing, I was tempted to start a scam selling "highest grade silicon ore." For $80 per pound, I would sell bags of beach sand.

Quartz is unique among the major rock-forming minerals in accepting only small amounts of other chemicals. (The objectionable iron in glass

Previous Page:
  Helices from left- and right-handed quartz crystals.

making sand comes from grains of nonquartz iron minerals.) Small amounts of iron, actually inside the quartz structure – plus some rearranged electrons – generate amethyst, treasured for its purple color. Traces of aluminum produce smoky quartz (called *cairngorm* in Scottish jewelry).

There is a huge variety of very fine-grained quartz materials. Flint, valuable for spear points and arrowheads, was the first resource to prompt underground mining. Among the less-valuable gemstones are agate and jasper; their structure is only now being understood. One kind of twinning in quartz, a mirror reflection known as *Brazil twinning*, can occur at the unit-cell level to form moganite. Agate and jasper contain various amounts of moganite. Peter Heaney, at Penn State, suggested that natural waters that contain only separate hydrated silicon dioxide molecules will grow clean quartz, but water containing silicon dioxide linked into polymers will grow moganite. I once took Heaney on a trip to sample jasper (sometimes called *jasperoid*) associated with Carlin-type gold deposits (#3). Almost all of the samples contained some moganite, which was a window into the chemistry of the ore-depositing water.

Despite its widespread natural abundance, quartz is officially classified as a hazardous substance. Many underground miners dealing with quartz-bearing ore died from lung disease. The sign at the highway turnoff to a Nevada ghost town reads: "Delamar, the widow maker."

Above:
   Gold ore from Delamar containing abundant quartz.

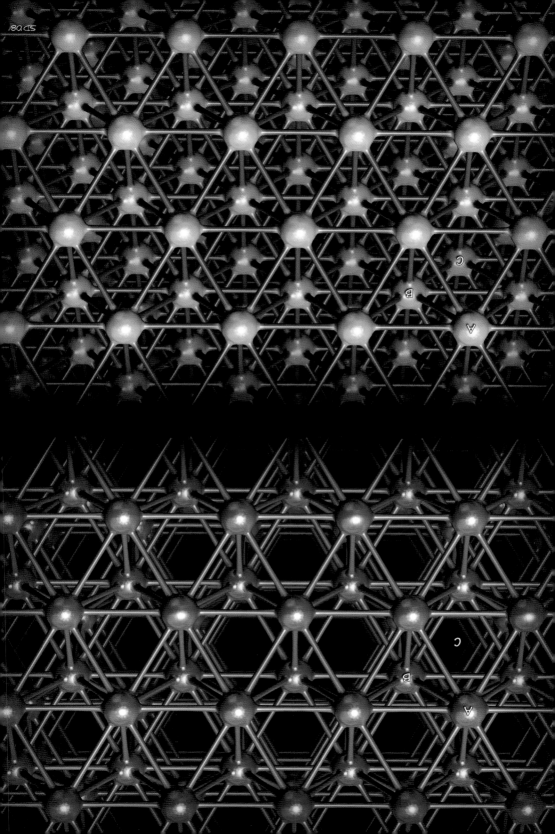

28   CLOSE-PACKED METALS

Many, but not all, simple metals arrange their atoms in closely packed structures, with each metal atom surrounded by twelve neighbors. However, there are two ways to achieve close packing, shown here by magnesium (above) and by gold (below).

If you look at the illustration for gold, three possible locations for the atoms are labeled (on the left) as A, B, and C. Reading into the page, the sequence is ABCABCABC...and so on. For magnesium, the same three locations are labeled, but atoms appear only at A and B, and the C location is vacant. The vacancy does not mean that the magnesium packing is more open; there are similar vacancies in the gold lattice but they do not line up when viewed from this angle.

Although magnesium and gold are both densely packed, magnesium has a hexagonal symmetry and gold is cubic. Although it is not a popularity contest, 22 metals share the hexagonal magnesium structure and 15 have the cubic gold arrangement.

Until recently, there was an outside chance that some other packing might be even tighter than the cubic and hexagonal close packing. A joke said that all mathematicians believed, and all physicists knew, that cubic and hexagonal were the closest possible packing. Because of a recent proof, mathematicians now also know that the two are the closest-packing arrangements. Although it seems counter-intuitive, most rock-forming minerals are closely packed oxygen ions, with the silicon, aluminum, magnesium, and iron ions propping the oxygens slightly apart.

In his 1939 book *The Nature of the Chemical Bond*, Linus Pauling explained that cubic and hexagonal arrangements were not the only possible variations on the ABC theme. We will see the sequence AABAAC in our illustration of the mineral erionite (#30).

Previous Page:
    Atomic structure of magnesium (above) and gold (below).

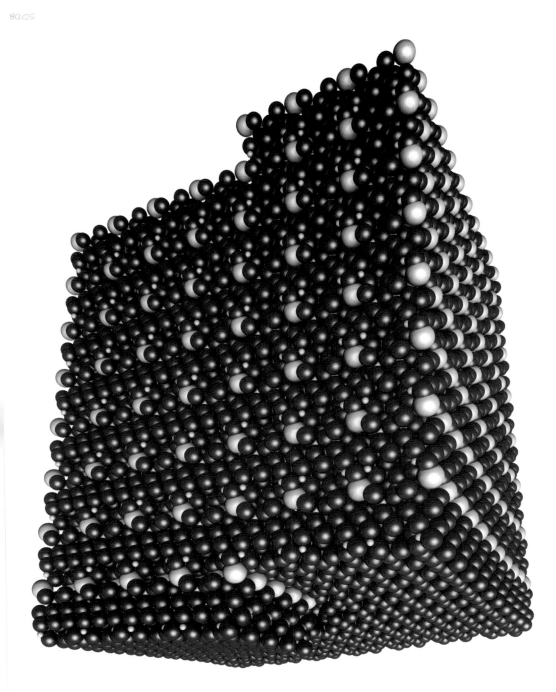

A considerable energy barrier usually must be surmounted in starting a new crystal. A crystal a few atoms across is unstable; somehow the system has to produce a nucleus of several thousand atoms before the crystal can grow. But this is not a rare occurrence – on some days water vapor trails from airplanes grow into cloud streams much larger than the original vapor trail.

A similar, but smaller, barrier should exist when a growing crystal completes one unit-cell layer and needs to start the next layer. However, some crystals seem to need no energy at all to initiate new layers. The barrier required to start a new layer would slow crystal growth by a factor of several thousand. The solution to the puzzle came when Burton, Carbrera, and Frank pointed out that a spiral-staircase step known as a *screw dislocation* allowed the crystal to grow without ever needing to start a new layer.

The illustration shows a simple example of a screw dislocation. The mineral chosen is biotite, black mica. The step is one unit cell high and it leads to a mica structure known as biotite 1M. The number one stands for the repeating single-cell height and M stands for *monoclinic*, the symmetry of the resulting crystal. (Other mica varieties have numbers like 2M and 3T.) The potassium atoms, between the mica sheets, allow you to trace out the spiral. The step on the upper surface ends at the screw dislocation. At the edges of the illustration, the structure looks like an ordinary mica crystal. The screw dislocation makes it unnecessary to start a new sheet.

Once we knew about screw dislocations, several things emerged. Not only does crystal growth revolve around screw dislocations; dissolving a crystal also is speeded up by a screw dislocation. Multiple screw dislocations can occur on a single growing crystal face. Left- and right-handed screw-dislocation pairs can occur. Growth promoted by a screw dislocation can generate almost-flat pyramids. These *vicinal faces* are sometimes visible in the light reflected from crystal faces.

Previous Page:
  Screw dislocation in biotite mica.

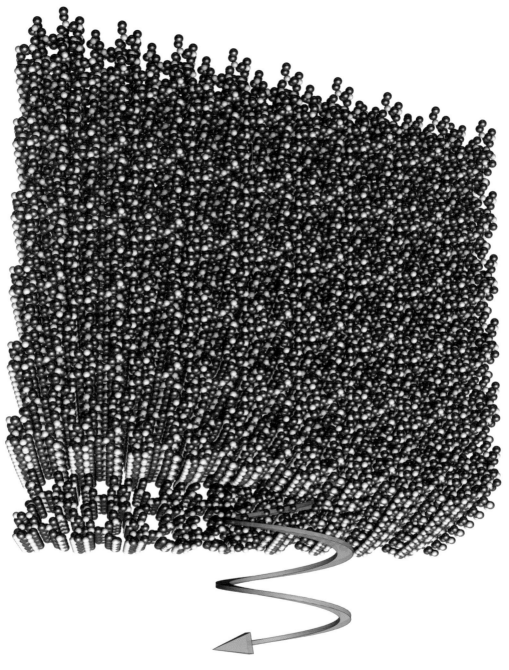

30 ERIONITE

From 1898 to 1956, the only known sample of the mineral erionite was a single specimen in the Harvard mineral collection. As a graduate student studying sediments in several Nevada basins, I found volcanic ash beds altered to several different minerals of the zeolite family. One of them I could not match to any obvious mineral, and the U.S. Geological Survey helped by saying that my x-ray diffraction data seemed to match the Harvard erionite specimen. The Harvard museum was kind enough to send me a small sample from the original erionite specimen, and it matched exactly to my unknown Nevada mineral. The known world supply of erionite jumped suddenly from a single museum specimen to millions of tons. I wrote up a new description of erionite that was published in the *American Mineralogist;* it was my first scientific paper. Erionite was my baby.

When we described close-packed metals (#28) the layer sequence in metals was described as AB,AB,AB for hexagonal and ABC,ABC,ABC for cubic. Erionite is made of silicon-oxygen six-rings and the horizontal six-rings are stacked vertically in the order AABAAC, AABAAC. How in the world does the next layer of six-rings copy a location far down in the stack? In the illustration, we speculate that a spiral dislocation might account for the regular repeats of the stacking order. Erionite was originally named for the Greek word for wool; the mineral usually (but not always) occurs as fine fibers. The obvious speculation is that there is one screw dislocation running down the length of each fiber.

Beginning in 1975, the whole character of the erionite story changed. Three villages carved out of volcanic ash in the Cappadocia region of Turkey had erionite present and a particularly deadly kind of cancer (mesothelioma) was the leading cause of death in the villages. Previously, mesothelioma was only known from exposure to asbestos fibers. Today, erionite is classified as more deadly than the worst forms of asbestos. My baby turned out to be a mass murderer. Like all parents of mass murderers, I could only say, "I don't understand it; erionite was such a nice quiet mineral."

Previous Page:
A screw dislocation repeating the six layers in erionite.

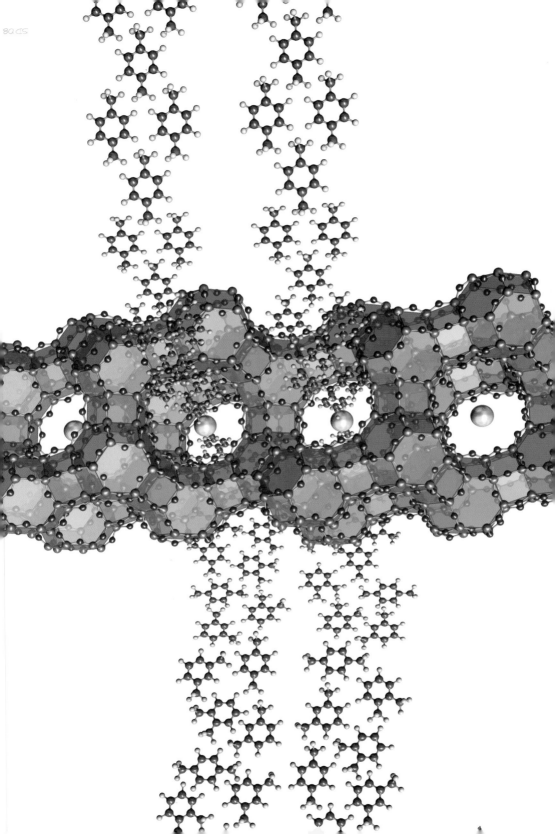

31 FAUJASITE

A catalyst speeds up a chemical reaction without itself being consumed. A good catalyst is similar to having the right ski wax; you slide downhill faster. However, no ski wax allows you to stand still and slide uphill. In chemistry, "downhill" is a combination of two things: 1) The inherent energy in each molecule, formally called the *Gibbs free energy*, and 2) the relative abundance of each type of molecule.

Platinum metal is a favorite catalyst, and accordingly industrial users have pushed platinum to twice the price of gold. If you use even tiny crumbs of platinum metal as a catalyst, most of the platinum atoms are inside the crumbs and only the atoms on the surface are doing their catalytic duty. A method for making each and every platinum atom active uses the *zeolite* mineral family. Natural zeolites consist of a rigid open lattice made of silicon, oxygen, and aluminum; roughly 100 different zeolite framework arrangements (natural and synthetic) are known.

In the illustration, the framework of the zeolite mineral faujasite is shown with oxygen in red, and purple for both aluminum and silicon. Gray shading emphasizes the faujasite framework. Single atoms of platinum (shiny white) replace the sodium ions that were originally in the large pores of the framework.

Schematically, at the top of the illustration we show an incoming feed of three kinds of xylene. Xylene consists of a carbon six-ring with two methyl (CH_3) groups attached. There are three possible xylene arrangements:

* With the two methyl groups on adjacent carbons
* With one carbon between the methyls
* With two carbons between the methyl groups

The third one, para-xylene, is valuable as a building block for the enormous group of polyester plastics. After the catalyst rearranges the methyl groups, para-xylene (the straight one) can exit the zeolite pores quickly and pour out toward the bottom of the page. In this particular example, the three xylene structures have rather similar inherent energies. The catalyst keeps

Previous Page:
Faujasite atomic structure, with platinum catalyzing xylene.

rearranging the three methyl groups and the chemical plant operator can separate and sell the para-xylene and recirculate the less-valuable xylenes back for another pass through the catalyst.

James Wei, then a professor at MIT, performed what initially might sound like a stupid stunt. He plugged up some of the pores in a catalyst for xylene. Production of the valuable para-xylene went up. Any para-xylene, once formed, could escape quickly through the few remaining open pores. The other forms of xylene would stick around and get hammered repeatedly until they took on the para-xylene geometry.

In addition to catalysts, zeolites have a lot of other uses: dehydration, adsorption, and ion exchange. In the early 1960s, I was helping the Shell research program because I had worked with a number of natural zeolite deposits. Shell had several chemists working on possible uses of zeolites and we were looking at all sorts of possibilities. Only one Shell chemist, Hans Benesi, was working on zeolites as potential catalysts. When Benesi would explain his work, the rest of us said, "Yes, Hans. That's nice, Hans." We didn't pay much attention. In contrast, the Mobil Oil research lab (before their merger with Exxon) realized that the big money would be in catalysts. Mobil ate our lunch. The industrial use of catalysts is now generating products worth a trillion dollars per year.

The illustration on the next page shows a close-up of a platinum atom inside a zeolite pore. After repeated heating cycles to remove accumulated carbon-rich "crud," the platinum-loaded zeolite gradually loses its internal chemical structure. The used zeolite is not discarded as waste, it is recycled to recover the platinum. I was an expert witness in one lawsuit after a platinum recycler recovered far less platinum than expected from a large shipment of used platinum catalyst. Fortunately, some of the samples that had been analyzed earlier were still available and we were able to show that someone had loaded the sample with additional pure platinum sponge.

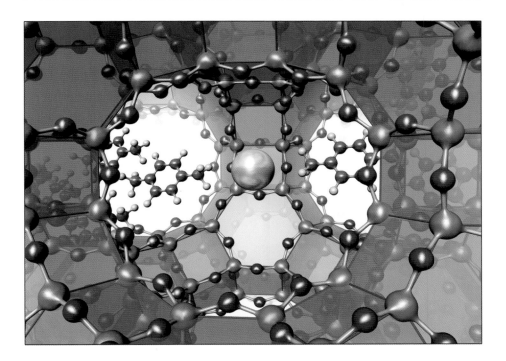

Above:
　　Enlarged view of a pore space in faujasite.

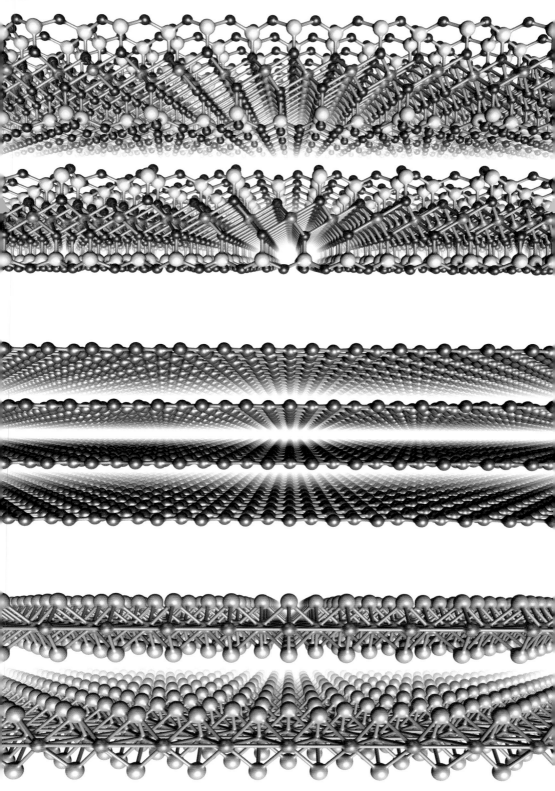

32 LUBRICANTS

Several minerals are used as lubricants. All of them have strong sheetlike crystal structures with weak bonds between the sheets. The weak bonds between the sheets are more a matter of physics than of chemistry. Most scientists use *van der Waals force* as sort of a wastebasket term for several different forces – attractive and repulsive – between nearby atoms. The weak bonding allows the sheets to slide past one another and the crystalline material can be used as a lubricant.

At the top of the illustration is the structure of molybdenum sulfide, MoS_2. It is the mineral molybdenite and is the source of virtually all of the world's molybdenum, used mainly in steel alloys. Small amounts of molybdenum sulfide powder are used as a lubricant, especially for use at high temperatures that would destroy oils, greases, graphite, and talc.

Graphite is shown in the middle of the page; it is almost pure carbon, either as a natural product or as a synthetic material. The "leads" in modern pencil leads are graphite; the product contains no chemical lead. Powdered graphite is sold as a lubricant; a typical application is blowing a small amount of graphite into a car door lock to make it work easily in cold weather. The strong "graphite" carbon fiber in sports equipment is produced by heating thin organic filaments to produce a strong lightweight fiber, which is very different from crystalline graphite.

At the bottom is the mineral talc. On the Mohs scale of hardness talc, the softest mineral, is given number 1 and diamond, the hardest, is number 10. The most familiar use of talc is as the main ingredient in talcum powder. An informal survey in a local drugstore showed that half of the baby-powder brands are made of natural talc, while the others are mainly cornstarch. Over the last 30 years there has been some controversy about whether small amounts of naturally occurring asbestos are present in commercial talc.

Previous Page:
   Mineral lubricants: sheet structures of molybdenite, graphite, and talc.

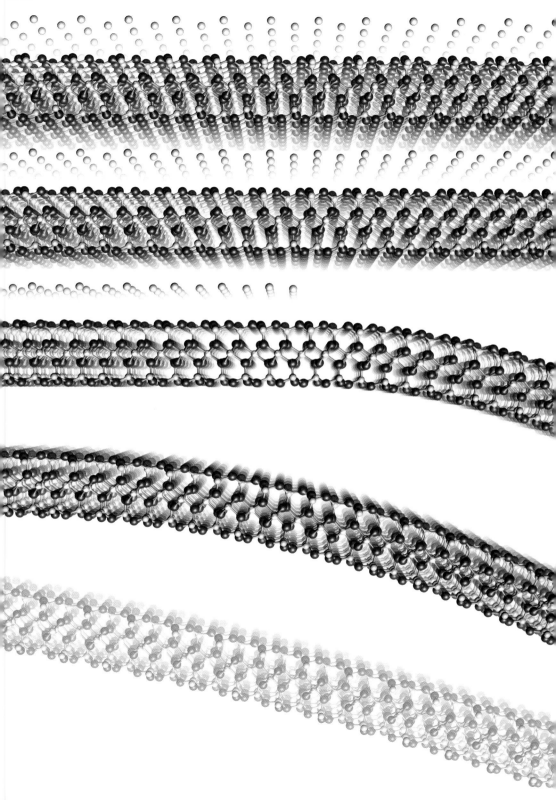

33   MONTMORILLONITE

Clay is a term used by geologists for the components of any extremely fine-grained sediment. Montmorillonite is a member of a subgroup of clay minerals that are properly called *smectite*. Unfortunately, *smectite* sounds like a dirty word; most geologists informally use *montmorillonite* as a name for the whole subgroup.

Montmorillonite, like the mica group (#29) and talc (#32), is composed of sheets. In montmorillonite, the sheets are bonded together by sodium or calcium, as shown in the bottom two layers of the illustration. (Oxygen is red, silicon is yellow, and aluminum has a reflective metallic treatment.) Water molecules can invade the space between the sheets and montmorillonite expands by about 40 percent when two water layers are added. Further water addition separates the sheets completely. The third layer from the bottom is shown peeling away and the interlayer sodium and calcium are blending in with the other minerals dissolved in the surrounding water. The fourth layer has separated completely and the fifth is wandering off into the water. Most of the commercial use of montmorillonite comes from the ability of the individual mineral sheets (about one nanometer thick) to separate in water.

Montmorillonite makes the mud for drilling oil wells, is important in some brands of kitty litter, is mixed with heavy oil to make grease, and is used as a low-leakage seal for waste-holding ponds. Montmorillonite turns up in toothpaste, cosmetics, and foods. It even figures in some alternative-health treatments.

Thick and relatively pure layers of montmorillonite are found in Wyoming and adjacent states; it is abundant and inexpensive. Natural, unmined, outcrops of almost pure montmorillonite look like a pile of gray popcorn. Geologists soon learn not to try to drive across them.

Previous Page:
Atomic view of montmorillonite layers separating in water.

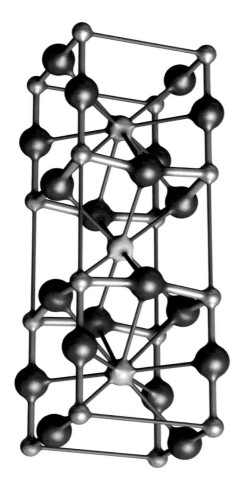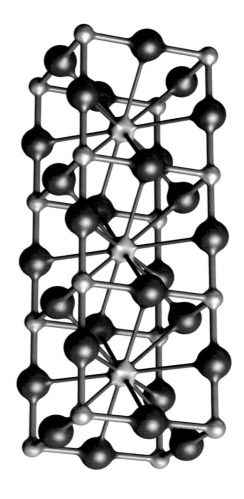

34 PEROVSKITE MORPH

In 1839, a new mineral was discovered in the Ural Mountains and named after a Russian mineralogist with the impressive name of Count Lev Aleksevitch von Perovski. For the next 140 years, perovskite had an obscure role as a minor rock-forming mineral. It was calcium titanate, $CaTiO_3$. Then, in the 1980s, perovskite emerged as a rock star.

Perovskite had two new, and newsworthy, roles:

* At high pressure, magnesium silicate ($MgSiO_3$) abandons its low-pressure pyroxene structure (#10) and dresses up in perovskite's structure. Earth's mantle deeper than 600 kilometers is inferred to be mostly magnesium silicate perovskite, making it the most abundant mineral on Earth.

* A different, and more radical, variation of the perovskite chemistry unexpectedly turned out to be a superconductor of electricity at much higher temperatures than previously known superconductors.

The example on the facing page shows how the original perovskite structure morphs into a different chemical composition with radically different properties. On the left is the original perovskite structure. Calcium is white, titanium is blue, and oxygen is red. The high-pressure magnesium silicate would be identical, with the coloring of magnesium as white, silicon as blue, and oxygen as red. On the left, the perovskite structure is cubic, and the illustration shows three identical unit cells stacked on top of one another.

The right-hand illustration converts the three unit cells into a new, larger unit. This is the celebrated high-temperature superconductor. In the center is an atom of yttrium in yellow; barium atoms (green) are at the center of the top and bottom cubes, copper atoms (copper-colored, what did you expect?) are at the cube corners, and red oxygens are along the cube edges. Notice that some of the oxygens are missing. This is not uncommon in minerals, and the missing oxygens play an essential role in the superconducting behavior.

Previous Page:
  Perovskite unit cells (left) morphed into a superconductor (right).

35 PEROVSKITE SUPERCONDUCTOR

The illustration on the facing page shows the full structure of the ytrrium-barium-copper-oxygen superconductor ($YBa_2Cu_3O_7$). At the top, middle, and bottom of the illustration are layers identified by the yellow yttrium atoms, accompanied by oxygen vacancies along the vertical cell edges. These divide the original cubic structure into a stack of nano-sandwiches. It is widely thought that the division into layers plays an important role in the superconducting behavior, but the mechanism is not fully understood. ("Not fully understood" is nerdish for "We haven't the foggiest idea how it works.")

The discovery of high-temperature superconductors was a real surprise. All of the earlier superconductors had been metals or metallic alloys; the perovskite version was essentially nonmetallic and did not conduct electricity at room temperature. The magnitude of the surprise: The superconductor was announced in 1986 and the discoverers were awarded the Nobel Prize for Physics the very next year. (Einstein had to wait 17 years for his Nobel Prize.)

The magnesium-silicate perovskite in Earth's lower mantle also contained an unexpected feature. At a depth of 600 to 700 kilometers in the mantle there is an abrupt increase in the speed of earthquake waves, probably caused by a change to a denser structure beneath. In the early 1980s, when the change was initially identified as a change from pyroxene above to perovskite below, Alexandra Navrotsky showed that the pressure-temperature boundary was not what we usually expect. Normally, higher-temperature transitions occur at higher pressures. This normal condition would boost the upward flow of warmer material and would also speed the sinking of colder oceanic plates. However, the Navrotsky insight restricts the easy convection through the 650-kilometer transition. Only the oldest and coldest down-going oceanic plates, as in the Japan Trench, and the buoyant upwelling plumes, like Hawaii, are able to move through the transition.

Previous Page:
Atomic structure of the perovskite superconductor.

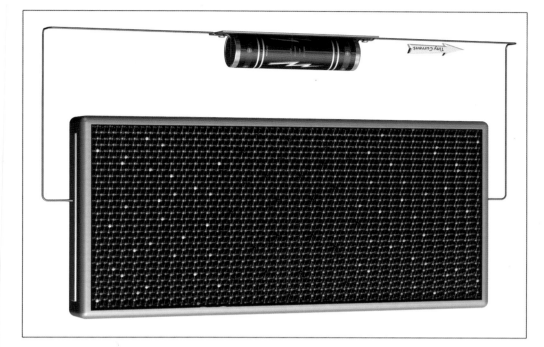

Tiny Current

Large Current

36

The simplest semiconductor device in this Age of Silicon is the diode. Somewhat incorrectly, a diode is called "a one-way valve for electricity." It is sometimes written that a diode "doesn't obey Ohm's law." Ohm's law says that the electric current through a resistor is a linear function of the voltage. I prefer to say that silicon diodes do obey Ohm's linear law, but they are cleverly arranged to go nonlinear for voltages much smaller than one volt.

The illustration shows the inside of a junction diode, made from silicon. The silicon in the right half of both illustrations has a small amount of boron (shown in white) substituted for the silicon. We have exaggerated the amount. If we showed the actual level of boron, there would be at most one boron showing in the whole illustration. Silicon carries four electrons in its outer shell, boron has three. Because the boron is short one electron, on the right side there is a net positive charge. On the left, phosphorous (shown in yellow) is introduced carrying five electrons, one more than silicon, so the net charge is negative.

The upper illustration has the negative pole of the battery connected to the phosphorous-substituted end. The two sides of the junction are in electrical contact and current can flow.

The lower illustration is identical, except the battery is reversed. In a zone next to the junction, each boron atom is supplied an extra electron and each phosphorous is stripped of its extra electron. These neutralized sites are shown in dark black. The junction behaves as an insulating layer and limits the flow of electrical current through the diode.

There are several useful variations on the diode theme. Where the electrons meet the gaps at the junction, visible light can be given off: a light-emitting diode or LED (#40). In the opposite case, light is allowed to shine on the junction and you have a photodiode or a solar cell. A diode can be a sensitive temperature detector. Adding a third connection to control the electrical field near the junction turns the diode into a transistor.

Previous Page:
  A silicon diode, shown at the atomic level.

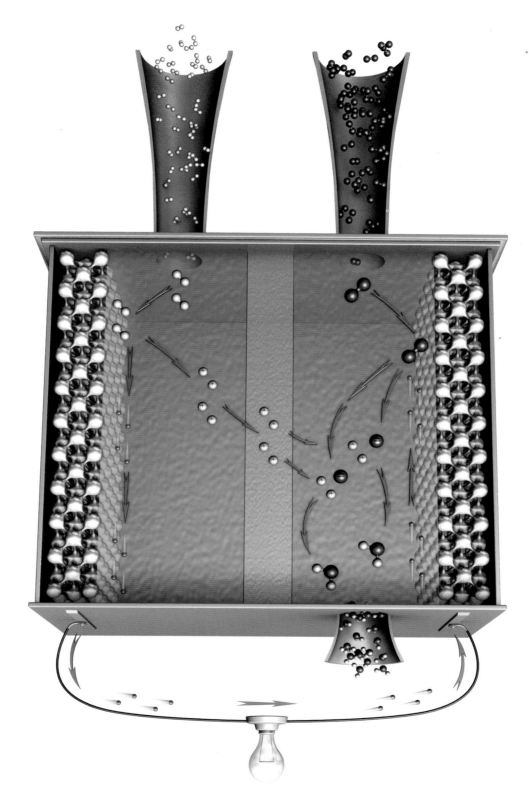

37

FUEL CELL

A fuel cell is a chemist's sweet dream and an engineer's nightmare. Our illustration shows the chemist's view. Although the fuel cell was discovered in 1839, the first important application of fuel cells came between 1969 and 1972. The Apollo moon missions used fuel cells to supply electric power and drinking water for the crew. (The near-disaster on the Apollo 13 mission was not from the fuel cell itself; it was from a heater in the liquid-oxygen tank.) The core of a fuel cell consists of a container of liquid divided into two halves by a plastic membrane. On either side are walls with electrically conducting surfaces.

On the right-hand side of the illustration, two molecules of hydrogen gas (H_2) enter at the bottom. They give up four electrons to the right-hand wall and four protons (H^+) remain in the liquid. Those four protons make their way, from right to left, through a tan-colored plastic sheet that selectively allows only protons to move. The four electrons travel up the wall and along the wire, from right to left, and do some useful work along the way as shown symbolically by the light bulb.

On the left-hand side, one molecule of oxygen gas (O_2) comes in at the bottom and picks up four electrons from the wall to form two oxygen ions (O^{--}). The two oxygen ions meet and greet the four protons from the plastic sheet to form two molecules of water (H_2O). It's lovely, it's clean; only electrical power and pure water come out.

So what makes the engineers unhappy? They have to bring the gas bubbles, the liquid, and the solid electrical conductors into intimate contact. Catalysts are required at the electrically conducting surfaces to speed up the chemistry and impurities in the fuel can disable the catalyst. Right now, and for the last 100 years, platinum metal has been the most effective known catalyst. Needless to say, platinum is expensive, more than $10,000 per pound. Worse, all the platinum produced from all the world's mines would suffice to build only 11,000 fuel-cell cars per year. Last year, world automobile production was 40,000,000 cars. The fuel-cell car isn't going anywhere until someone discovers an alternative catalyst.

Previous Page:
  Chemical view of a fuel cell.

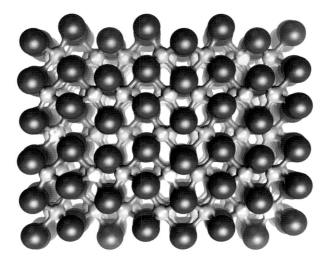

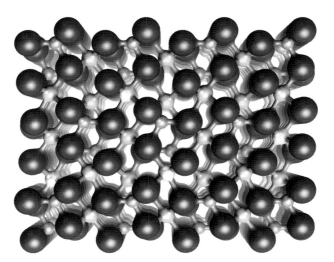

38   LASER CRYSTALS

A *laser* (for "light amplification by stimulated emission of radiation") is a material that can take in energy from one source and turn out a coherent light beam. The theory behind the laser goes back to Einstein in 1916 and 1917. The first practical gadget, in 1953, amplified microwaves and was called a *maser*. The race to produce the same effect in visible light was won in 1960 using a synthetic ruby crystal, as shown on the next page. Since then, an enormous range of solids, liquids, and gases has been used in lasers, and they have been powered by light, by electrical current, and by chemical reactions.

In a laser, external energy boosts electrons from one allowed energy level to a higher level, where they reside temporarily. (In different materials, "temporarily" can be anywhere from microseconds to minutes.) There follows sort of a groupthink decision for the electrons to drop back to a lower level, emitting a beam of laser light.

A solid laser material needs to be stable, clear, readily polished, heat resistant, and insoluble in water, and it must have interesting interactions with light. The same requirements apply to gemstones. Once an interesting natural material has been identified, synthetic crystals with modified chemical compositions are usually produced with enhanced properties. Three natural gemstones are shown in the illustration; each of them has had a second career as a laser material.

The top crystal in the illustration is alexandrite. The heavy magic of alexandrite as a gemstone is a color change from green in daylight to raspberry red in candlelight (or incandescent light). Alexandrite was discovered in Russia and named in honor of Tsar Alexander II. It is a variety of chrysoberyl (Al_2BeO_4) with about 0.1 percent of chromium replacing the aluminum. Among the producers of alexandrite for lasers is a division of Northrop Grumman, the defense contractor. However, the largest use of alexandrite lasers is in dermatology – erasing tattoos and removing unwanted hair growth.

The middle crystal is ruby, which is the mineral corundum (Al_2O_3) with 0.1 percent of chromium replacing the aluminum. The color response

Previous Page:
Laser crystals, from top to bottom: alexandrite, ruby, and garnet.

of chromium is determined in part by the surrounding atoms. Chromium makes rubies red, but it also generates the green of emerald and imperial jade. Historically, the largest original source of high-quality rubies was in Burma (now Myanmar). During World War II, many of the American soldiers in Burma were aware that the area was famous for its rubies; the locals broke the red glass tail lights out of Jeeps, cut them into faceted stones, and sold them back to the Americans.

In 1902, August Verneuil started production of synthetic rubies, with the same crystal structure and chemistry of natural rubies. The Verneuil process used a mechanical salt shaker to sprinkle the aluminum and chromium oxides onto an oxygen-hydrogen flame. The result was a single crystal a bit larger than your thumb. Unfortunately, there were bubbles in the crystal and the rubies were readily distinguished from natural crystals. A superior method for growing crystals was invented by Jan Czochralski in 1916. A container of the material was melted and then a single seed crystal was gradually pulled up out of the melt. Cooling would grow a large, comparatively clean crystal. (People who can't spell – or pronounce – Czochralski call the products *pulled crystals*.)

The garnet structure, shown at the bottom of the illustration on the previous page, is not disordered; it just looks messy. Actually the garnet unit cell contains a large number of symmetry elements that repeat a single oxygen ion into 96 different locations. Clear natural garnets, typically red, are not rare and they have been used in jewelry since ancient times. The garnet in the illustration has come a long way from its original chemistry. Only the oxygen (red) and the symmetry remain. Silicon, aluminum, and magnesium have been replaced by gallium (yellow) and gadolinium (purple). The result is usually called GGG for gadolinium gallium garnet. The use of GGG as a laser crystal is made possible by adding 1 percent of neodymium. The Northrop Grumman Web site describes the neodymium GGG only as a "military laser." Other Internet postings suggest that a Nd:GGG laser can quickly burn a hole through a one-inch thick steel plate.

Above:

    The original ruby laser consisted of a spiral flash bulb wrapped around a rod of synthetic ruby crystal. When the flash bulb illuminated, the ruby responded by emitting a strong beam of red light.

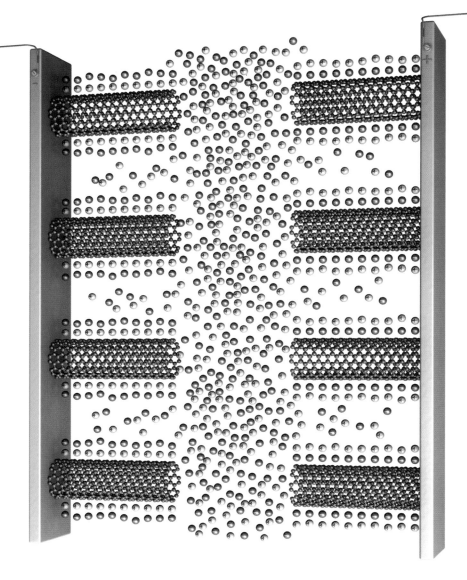

39 SUPERCAPACITOR

Capacitors, which store electric charge quickly, used to be distinct from batteries, which are slower in taking up and delivering electricity. Recently, several devices in between capacitors and batteries have appeared. This section explains one of those intermediate devices. At the moment, the same system exists under several names: supercapacitors, ultracapacitors, double-layer capacitors, and even gold capacitors. Structurally, a capacitor consists of two metal sheets separated by an insulating layer. Although it isn't intuitively obvious, making the insulating layer thinner increases the electrical storage of the capacitor. The size of a capacitor is measured in farads. A one-farad capacitor charged with one volt can deliver a one-ampere current for one second.

The double-layer supercapacitor lets nature do one of the difficult manufacturing steps. As early as 1879, it was recognized that the dissolved ions in a liquid (such as water) would form a double electrical layer adjacent to an electrically charged solid surface. Around 1990, the double layer began to be used in capacitors. The effective thickness of the insulating layer was a single layer of ions adsorbed on the solid surface. The size of capacitors jumped from picofarads and microfarads to thousands of farads. Maxwell Technologies is currently marketing a 3000-farad capacitor the size of a pint milk bottle.

The price paid for the huge electrical capacity is a low voltage. Water starts to break down into hydrogen and oxygen gas at one volt. Substituting acrylonitrile for water raises the voltage to 2.5 volts, but the capacitor has to be sealed up. Acrylonitrile can cause a wide variety of things, from fire to cancer.

Our illustration shows the next-generation supercapacitor dream. Carbon nanofibers (#8), with huge surface areas because of their small size, support the double layers. At least half a dozen university labs and several commercial labs are racing to deliver the first commercial nanofiber capacitor.

Warning: Some capacitors can deliver lethal electrical shocks. Unless you have a bachelor's degree (or higher) in electrical engineering, do not conduct experiments in your garage. Small capacitors – the size of a pencil eraser or smaller – in electronic circuits generally are not dangerous.

Previous Page:
An engineering goal: A double-layer capacitor based on carbon nanotubes.

40 EPITAXIAL GROWTH

A crystal that takes its orientation from growth on a pre-existing crystal is called *epitaxial*. Although inorganic epitaxial crystals occur naturally, they are not at all common. In the organic world, solid structures from seashells to our bones and teeth are controlled by crystal growth on optimized organic substrates. The illustration shows controlled epitaxial growth used to produce blue light-emitting diodes.

The first synthetic epitaxial material that I remember reading about was in 1945. During World War II, an optical gun sight was produced from large clear natural crystals of calcite. However, optical-quality calcite was in short supply. Sodium nitrate, which has the same crystal structure and optical properties as calcite, was melted; a sheet of natural mica was placed on the melt and the assembly was allowed to cool. A single large crystal of sodium nitrate grew epitaxially on the mica. Big surprise: sodium nitrate and mica didn't seem to me to be closely related. Bernard Vonnegut's ice crystals grown on silver iodide (#2) are epitaxial.

Epitaxial growth is widely used in the semiconductor industry for all sorts of devices. We have selected the blue light-emitting diode (LED) as an example. Red LEDs were first announced in 1962 and during the 1980s a whole variety of orange, yellow, and green LEDs became available. A blue LED did not appear until 1993, following a breakthrough by Shuji Nakamura.

The blue LED involves an epitaxial junction between gallium nitride and gallium-indium nitride. Although there are several different commercial methods for generating epitaxial growth, we selected chemical vapor deposition for the illustration. Initially, two gases (ammonia and trimethyl gallium) are present and the lower half of the crystal is gallium nitride (purple and blue). For the upper half of the crystal, trimethyl indium is added to the gases and indium atoms (yellow) are incorporated to produce gallium-indium nitride. Notice that the crystal structure is continuous and uninterrupted across the junction. Seen escaping into the gas at the top of the page are byproduct molecules of methane (CH_4).

Previous Page:
Epitaxial growth of an iridium–gallium nitride crystal on top of a gallium nitride crystal to form a blue light-emitting diode (LED).

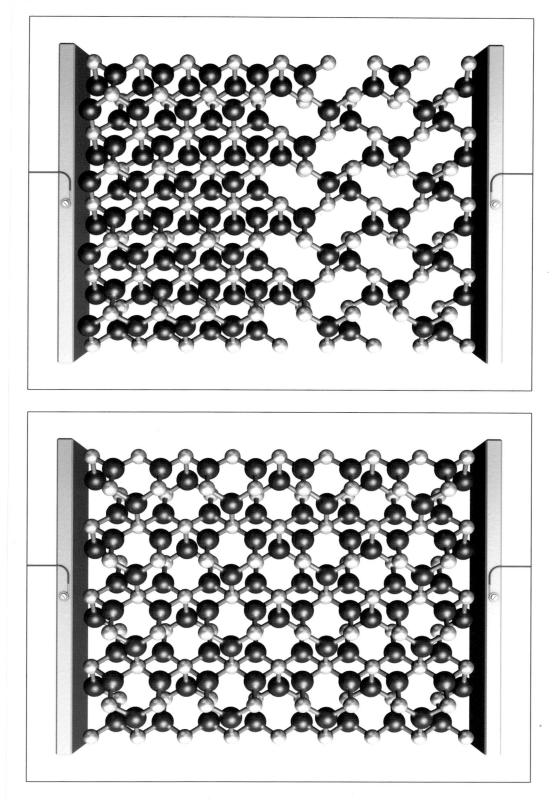

41

In 1971, Leon Chua at U.C. Berkeley pointed out that resistors, capacitors, and inductors (coils) should be joined by a fourth entity, which he named a *memristor*, but no actual memristors seemed to exist. In 2008, Hewlett-Packard labs announced that they had built working memristors.

In #36, Silicon Diode, we explained that a diode has a nonlinear behavior because electrons move to create insulating and electrically conducting zones. In contrast, in a memristor whole atoms are moved around to change insulators into conductors. Moving atoms is a blessing and a curse. When the electric power is turned off, the atoms in the memristor just sit still. When the power comes back on the memristors retain their original setting – that's instant boot – which is the blessing. The name *memristor* stands for "memory resistor." In contrast, the electron configuration in a diode leaks out after the power goes off.

The curse comes because the memristor takes a while to move atoms. An analyst who dissected the 2008 announcement claimed that the Hewlett-Packard memristor took one whole second to respond. To computer engineers, one second is forever. To minimize the atom-moving time, memristors have to be thin in the direction of the electric current. The Hewlett-Packard gadget is about 15 atoms thick; a memristor is inherently a nanoscale device.

The Hewlett-Packard active material is titanium dioxide. There are three different titanium-dioxide mineral structures: anatase, rutile, and brookite. The illustration on the opposite page shows the rutile structure because rutile often has oxygen vacancies. The unique axis of our illustrated rutile structure points left to right across the page. The large red spheres are oxygen. Titanium (plus a small amount of iron) is shown as the small white spheres. The upper illustration shows a memristor, with its oxygen vacancies evenly distributed. Electron transport is promoted by the oxygen deficiencies and the upper memristor will conduct electricity.

In the lower illustration, the vacancies have migrated to the left. On the right is defect-free rutile, which is a good electrical insulator; the sandwich does not conduct electricity. The state of the memristor can be determined by sending in a high-frequency alternating current to measure the electrical conductivity without resetting the memristor.

Previous Page:
  A memristor, shown at the atomic level, in both on and off states.

More than 1,000 years ago, the Chinese used lodestone (natural magnetite) on a float in a cup of water as the first magnetic compass. That's the "first" for people; the birds and the bees had used built-in magnetic compasses for millions of years. About 50 different species, from bacteria to migratory animals, can detect Earth's magnetic field.

The illustration shows that the crystal structure of magnetite contains iron in two different settings. Iron is magnetic because unpaired electron orbits carry an electrical current, like the current in an electromagnet coil. One location for iron, with four neighboring oxygen ions, is shown by the single arrow indicating a magnetic contribution downward and to the left. The other iron location (with six oxygen neighbors) is flagged by two arrows showing a magnetic field upward and to the right. Magnetite is magnetic only because the sixfold irons are twice as numerous as the fourfold irons. All magnetic materials are attracted to a magnet, but only some of them (including magnetite) are permanent magnets, they retain a magnetic field.

Biological magnetic sensors are a fascinating story. I happened to participate in the opening chapter, which involved honeybees. My task was to make an x-ray shadowgraph of a dried bee, similar to what a radiologist does in the hospital. The difference was that "hard" medical x-rays go through meat and are stopped by bone. For the bees, I needed to use a "soft" x-ray wavelength that was strongly absorbed by iron.

The crystallographic x-ray diffraction machine could deliver the appropriate soft x-rays but safety interlocks shut down the machine if anyone opened one of the x-ray ports without having one of the manufacturer's machines mounted in front of the port. In order to use my homemade bee machine, I stuffed paper wads in the switches to disable the safety interlocks. I had to be really, really careful. Those "soft" x-rays deliver enormous surface doses, a quick route to skin cancer or eye cataracts. It worked; on my second try I got the exposure correct and two iron-rich dots showed up at the front of the bee's abdomen. I quickly retired from my short career as a bee radiologist – no skin cancer, no cataracts.

Previous Page:
Magnetite (Fe_3O_4) with arrows showing iron–atom magnetic orientations.

RARE EARTH MAGNETS

Around 1980, powerful and relatively inexpensive permanent magnets became available. The new magnets contain *rare earths*, a family of chemical elements (atomic numbers 57 to 71) with useful optical and magnetic properties.

Rare earths are not exactly "rare." Neodymium (shown in purple in the illustration) is more abundant than lead in Earth's crust. Neodymium is the major rare earth in the most successful of the new magnets, which also contain iron (light brown) and boron (dark blue). The magnetic material is produced by mixing the dry powdered ingredients and heating the mixture in an atmosphere-controlled furnace. In the lab, this is called a "Shake 'n Bake" recipe. (Thank you, Kraft Foods.)

The neodymium-iron-boron crystal structure is not easy to determine; crystals big enough for single-crystal x-ray study do not come out of the oven. The published crystal structure, shown in the illustration, is fascinating because there are neodymium-boron layers that resemble the layering in the perovskite superconductor (#35). There is a concern: The published structure may be an exception to the crystal-structure rules set out by Linus Pauling.

The high strength of the neodymium magnets is a hazard. A child who swallows one button-size magnet is not at risk, but the attraction between two swallowed magnets can punch a hole in the digestive tract. People with implanted heart pacemakers or defibrillators are advised to stay away from strong magnetic fields. And, as I can report, the larger neodymium magnets can inflict damage. I bought two large magnets to set up a tabletop magnetic test range. When I was putting the magnets away, I placed one magnet on top of the bookcase and picked up the second magnet off my workbench. The first magnet flew off the bookcase and trapped one of my fingertips in a magnetic sandwich. I wasn't strong enough to pull the two magnets apart. All the tools I grabbed from the workbench were iron, I couldn't control them next to the magnets. I was ready to call 911, but in desperation I was able to pull my fingertip straight out from between the magnets.

Previous Page
   Published atomic structure for a strong neodymium–boron–iron
   permanent magnet.

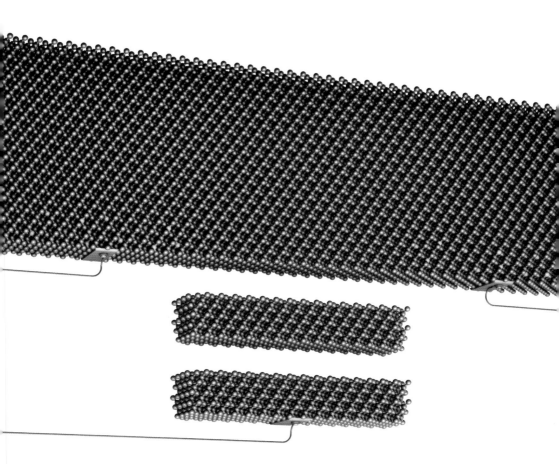

44 FLASH MEMORY

The modern type of flash drive, first marketed in 1989, has become a ten-billion-dollar-per-year market. Flash memory evolved from the field effect transistor (FET). An FET uses the electric field from one external connection to control the electron flow through semiconductor silicon from a source to a drain. For flash memory, an additional slice of silicon is included between the control and the semiconductor path. This additional slice isn't electrically connected to anything. (The white spaces in the illustration are electrical insulators.) Normally, the FET and the flash memory use voltages of one to three volts. However, if a twelve-volt difference is applied, electrons can move into – or out of – the independent layer by quantum-mechanical tunneling through the insulator.

The electrical charge trapped in the insulated layer overrides the voltage on the topmost control layer and the modified FET then has a memory, a *flash memory*. The trapped, insulated electrical charge leaks out very slowly; it takes more than ten years to escape. Don't archive your children's baby pictures on a flash drive; the pictures may be gone before they graduate from high school.

I love the USB flash memory sticks; I can't keep myself from buying them. What used to be called a *sneaker net*, implemented with floppy disks, is replaced by flash memory sticks. One of my treasured possessions is an ultra-miniature flash drive, about 15 by 30 millimeters. When I give an out-of-town lecture, a backup copy of my PowerPoint slides is on the teensy flash drive, clipped to my neck wallet along with my boarding pass and passport.

A new stunt, made possible by flash memory, is packaging a computer's entire operating system and memory storage on a large flash drive. You can reboot a different computer from the flash drive and it has the familiar warm fuzzy feel of your home computer. Stephen Deffeyes, my coauthor, assembled a flash drive for me with DOS and BASIC. I boot a laptop from his flash drive, the computer thinks the year is 1983, and I can run my homemade BASIC programs.

Previous Page:
Atomic view of a silicon flash-memory circuit.

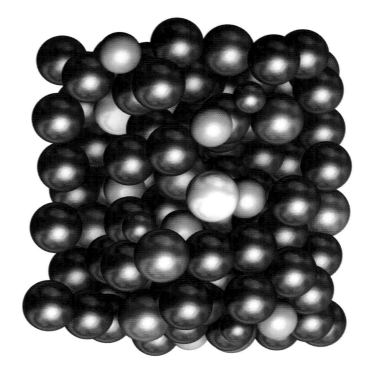

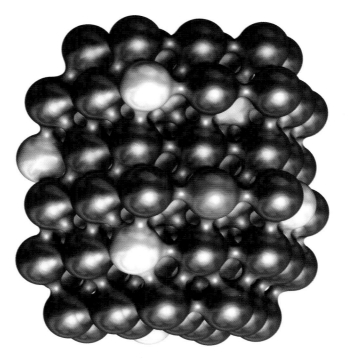

45

When cooled slowly, virtually all molten substances will solidify into well-ordered crystals. However, for some melts "slowly" can mean more than a year. If cooled rapidly, these slow-to-crystallize substances solidify into a liquidlike disorganized solid. Window glass and camera lenses are familiar examples. Metals crystallize even more rapidly, but in 1957 Sol Duwez at Cal Tech cooled a small splat of molten gold-silicon alloy on a copper plate, pre-cooled with liquid nitrogen, and obtained a disordered metallic glass.

Commercial production involves squirting molten metal on a rotating chilled drum and peeling off a ribbon of metallic glass partway around the drum. The metallic glass ribbon comes off at 27 meters per second (about 60 miles per hour) and is wound onto a take-up spool. I have a nightmare about a failure of the take-up spool and the room filling with razor-sharp metallic ribbon. Despite the manufacturer's warning, I have cut my fingers on the ribbon corners.

Metallic glasses found markets because of two unique properties:

* Dislocations are the major way that a crystalline solid deforms; an example is calcite twinning (#25). Metallic glasses are totally disordered and dislocations cannot exist. As a result, metallic glasses are very strong. One recent development is metallic glass heads for golf clubs.

* In the absence of crystalline boundaries, metallic glasses require very little energy to reverse their magnetic orientation. Energy-efficient magnetic cores for electric power transformers have been the largest commercial market for metallic glasses.

The upper example in the illustration is an example of a crystalline magnetically hard alloy used in computer disk drives, composed of cobalt (78 percent), chromium, and tantalum. The lower example is a magnetically soft metallic glass composed of cobalt (75 percent), iron, silicon, nickel, and boron. The important contrast between their magnetic properties comes from the atomic structure and not from the chemistry.

Previous Page:
  (Above) crystalline cobalt alloy, (below) metallic glass.

46   Spinodal Decomposition

Almost everyone has used spinodal decomposition: Dissolve powdered Jell-O (thank you again, Kraft Foods) in hot water to make a liquid. Let it cool and it segregates into a weak – but solid – network of gelatin with the water, flavor, and color trapped inside the gelatin network.

The word *spinodal* comes from a line with a curved bump or hump, a "spine," on a temperature versus composition graph. Underneath the curved line, a material of mixed composition can separate into regions of different composition. (The decomposition could have been called *bumpodal* or *humpodal*, but *spinodal* sounded classier.)

Although spinodal decomposition is not an everyday phrase, lots of examples turn up. For centuries, blacksmiths have manipulated carbon dissolved in iron to produce "tempered" steel tools. Virtually all the aluminum that we use is alloyed with one or more other metals and then "aged" or "precipitation strengthened." The tactic, in steel or aluminum, is to generate fine grains with a different composition that restrict the motion of dislocations through the bulk metal. (See #25 and #29 for examples of dislocations.)

And rocks – you didn't think that I would ignore rocks. The most spectacular mineral example is the spinodal decomposition of sodium-potassium feldspar. At high temperatures, sodium and potassium ions are completely interchangeable, and rapid cooling, such as in a surface lava flow, will preserve the disordered sodium-potassium feldspar. With slower cooling, the sodium and potassium segregate into separate domains; the result is called a *perthite*. In the illustration, the sodium ions (white) are gathered into a two-inch-wide band running from the upper left to lower right. In the actual feldspar, the compositional boundary is gradational.

Perthites exist on all scales. With slower and slower cooling rates, the thickness of the perthite layers range from those detected by x-ray diffraction (as in the illustration), through sizes comparable to the wavelength of visible light (moonstone, a gemstone), to thicknesses seen through a microscope (microperthite), to the millimeter-thick perthite bands in my daughter's granite kitchen counters. You can prepare a gelatin sauce, admire the perthite, and realize that they both have the same physics.

Previous Page:
    Feldspar crystal showing perthite zoning between sodium and potassium.

"If life hands you a lemon, make lemonade."

— Anonymous

Pipe carrying natural gas from deep, and hot, wells often gets plugged up by a white crystalline substance. The white substance turns out to be made up of small hydrocarbon molecules that have the same structure as tiny portions of the diamond lattice.

Deeper than about 15,000 feet (4.5 kilometers) the temperature is high enough to break down crude oil into natural gas. The last surviving oil molecules bear a resemblance either to the structure of graphite or of diamond. The graphitelike molecules are asphalt. The diamondlike structures are the white crystalline crud that clogs the pipe.

The illustration shows the three smallest diamondlike molecules, highlighted against a background of the diamond structure:

Adamantane, at the top, has 10 carbon atoms and 16 hydrogen atoms. Adamantane, named for the Greek word for diamond, was first identified in crude oil from Czechoslovakia. A chemical company, Lachema, in the present-day Czech Republic can supply adamantane at rates of tons per month.

Diamantane, in the middle, has 14 carbons and 20 hydrogens. Lachema claims a production capacity of 10 kilograms (22 pounds) per month.

Triamantane, the bottom molecule, with 18 carbons and 24 hydrogens, is available in trial quantities from Chevron.

Scientists are busy trying to write lemonade recipes for these lemons. One example of a useful molecule is adamantane with one hydrogen replaced by an amine (NH_2) group; it is a drug used in the treatment of Type A influenza. Lachema has on the shelf a dozen other adamantane derivatives waiting for applications.

There are even larger molecules in the series. Chevron claims to have identified diamondlike molecules with up to 50 carbon atoms. Chevron is not alone in the search; patents have been assigned to ExxonMobil and to Shell.

Previous Page:
   Three different hydrocarbon molecules, highlighted against the diamond structure.

48   PENROSE TILING

The ancient Greek geeks knew that only three regular polygons (with equal sides and angles) could be used to tile a floor: triangles, squares, and hexagons. In addition, there are mixtures. My bathroom floor is tiled with half squares and half octagons. But none of the schemes involved a fivefold symmetry. Classical crystallography, based on repeating patterns, was a mathematically complete theory. Nothing could be added.

New fivefold nonrepeating tile patterns were introduced in 1973 and 1974 by Roger Penrose, a mathematical physicist. Penrose patented his system and successfully pursued a lawsuit against Kimberly-Clark for embossing a Penrose tiling on their toilet paper.

The Penrose tiles shown on the opposite page consist of a fat rhomb and a skinny rhomb. In addition, there are matching rules, enforced by the circles and squares on the tiles. The pattern never repeats. There are more fat rhombs than skinny rhombs; the big surprise is that the ratio of the number of fat rhombs to skinny rhombs is the classical Greek golden mean or golden ratio. The number is 1.6180..., the only value of x that satisfies the equation $x = 1 + (1 / x)$. It's a number that goes on for an infinite number of decimal places without repeating.

In 2007, Peter Lu (my former student) and Paul Steinhardt described 500-year-old Islamic mosaics using the same principle that Penrose developed. If these mosaics, in buildings from Uzbekistan to India, had been recognized in 1973, Kimberly-Clark could have convinced the court to throw out the Penrose patent.

Almost all crystallographers were startled in 1984 when Dan Schechtman and his colleagues published observations on a metallic alloy with a fivefold symmetry that was forbidden in classical crystallography. At the time, Linus Pauling argued that Schechtman's samples were simply twinned classical crystals. In the same year, Paul Steinhardt published a theoretical argument that the fivefold symmetry was projected down to us from a six-dimensional space and named the new materials *quasicrystals*.

Previous Page:
A Penrose-tiled floor.

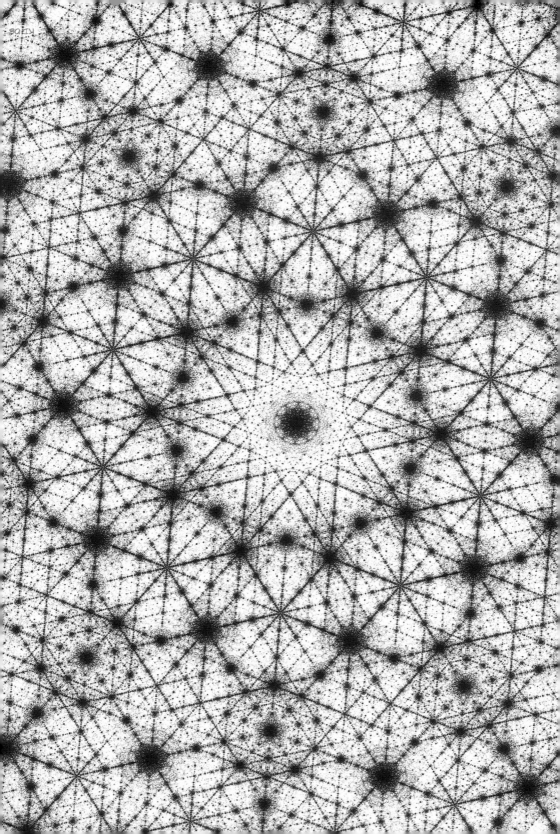

Sending a beam of x-rays through an ordinary crystal produces an array of weaker x-ray beams that reflect the symmetry and structure of the crystal. The Penrose tiling on the previous page does not repeat with a regular spacing. Intuitively, the tiling would not be expected to generate a sharp diffraction pattern. To the surprise of the few people who were paying attention, the Penrose tiling and three-dimensional quasicrystals (#50) generated sharp diffraction peaks. However, there was a difference. Ordinary crystals generated an array of x-ray beams with little or no energy between the various beams (#22). The Penrose tiling gave strong x-ray peaks, with weaker peaks in between, and tiny peaks between those. The x-ray pattern was a fractal, with smaller units showing structures similar to the larger units. The Penrose diffraction pattern has structure on all length scales.

The illustration shows the diffraction pattern from an opaque sheet with a small hole drilled at each corner of the tiles that are shown in #48. If x-rays or visible light were passed through the holes, diffraction would generate the pattern shown in the illustration. This is a computed pattern rather than an experimentally generated pattern. In the bad old days, the computation would have involved quantum mechanics and mathematical Fourier transforms. However, in the first two chapters of his 1985 book *QED*, Richard Feynman explains his sum-over-histories approach. His explanation was so straightforward and so simple that I not only understood, I was able to use his method for my own purposes. My homemade computer program that generated this illustration consists of only 16 lines of code (see p. 132).

The illustration shows the arrival locations for the x-ray energy as dark dots on a white background. That's a salute to the early days of x-ray diffraction when the x-ray energy exposed black dots on photographic film. The Penrose tiling in the previous section has a fivefold symmetry, but this illustration shows tenfold symmetry. That doubling happens because diffraction (real or computed) adds a center of symmetry.

Previous Page:
  Diffraction pattern from a Penrose tiling.

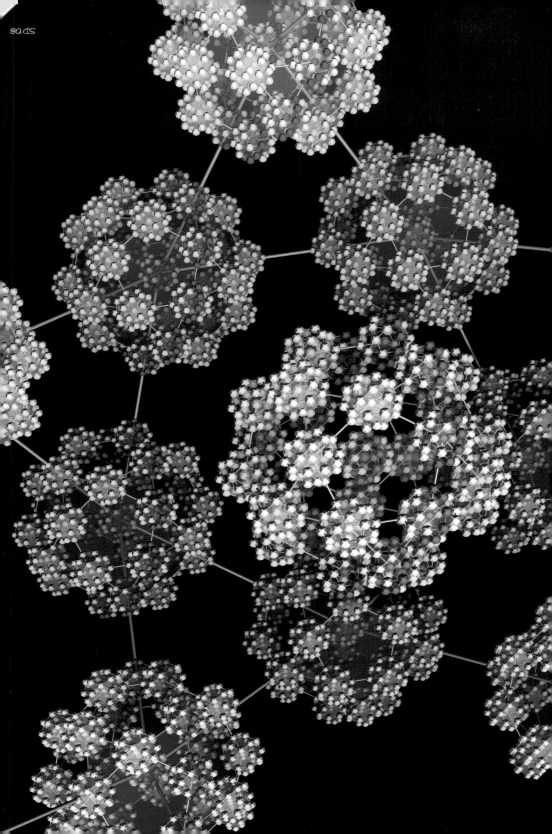

Within two years of Laue's discovery of x-ray diffraction, structures of conventional crystals were being deciphered. Twenty-three years passed from the discovery of quasicrystals to the first determined structure, and even then the structure contains some flexibility for the arrangement of atoms.

In 2007, Hiroyuki Takaura and his colleagues published a structure for a metallic alloy containing 5.7 cadmium atoms for each atom of ytterbium. (Ytterbium is one of four chemical elements named for the same village in Sweden; there is an unusual geologic deposit next door.) There were several advantages to working with that particular quasicrystal:

* It contained only two chemical elements, instead of three metallic elements as in most known quasicrystals.

* X-rays respond to the electrons in atoms and there is a large contrast between cadmium with 48 electrons and ytterbium with 70.

* Several classical crystals (not quasi-) have compositions near the cadmium-ytterbium quasicrystal. These classical crystals give hints about arrangements that might exist inside the quasicrystal.

The normal crystals close to the quasicrystal composition tend to have very large and complex unit cells. In one sense, a quasicrystal has an infinitely large unit cell. The illustration on the opposite page shows the "deflation" property in which smaller quasicrystal units repeat the organization of larger units.

Quasicrystals have come to the kitchen. A French company, Sitram, produces quasicrystal-coated skillets under the trade name Cybernox. We have used the pans. They are stick-resistant at high temperatures and the coating is hard enough to be scratchproof.

Discoveries that redefine a field, like quasicrystals, can arrive 100 years apart. It's not a good strategy to drum your fingers on the table while waiting for the Next Big Thing. There is lots of work to be done, using existing techniques, on biological structures, on materials for a quantum computer, and on energy-conserving composite materials.

Previous Page:
Proposed atomic structure of a cadmium-yttrium quasicrystal.

NOTES

Introduction – The Reality of Atoms

From 1905 until 1910, Albert Einstein published several papers using different routes to determine Avogadro's number (the number of atoms in a gram mole). Avogadro's number determines the size, and indirectly the reality, of atoms and molecules.

Einstein used several different kinds of observational data in his papers:

* The osmotic pressure of sugar solutions (see the following note),
* The random motion of micron-size particles in a liquid (Brownian motion),
* Opalescence of fluids near a critical point, and
* Light scattering in the atmosphere (blue sky).

All of the routes gave the same value for Avogadro's number, to within a few percent. The importance of the multiple approaches was not to improve the accuracy; the point was that the number was inescapable. Atoms exist.

A famous examination question goes: "The sky is blue. Compute Avogadro's number." The student goes into sticker shock because no numbers are supplied. For a rough exam-room calculation, we need only two numbers: The volume occupied by one gram-mole of gas at standard temperature and pressure, 22.4 liters, and the wavelength of blue light, 400 nanometers (red light is 700 nanometers).

The core of the calculation is the statistical uncertainty in the number of molecules in a teensy cube of gas; a cube the size of a wavelength of blue light. The gas molecules are rattling around at random, and the statistical uncertainty is roughly equal to the square root of the average number of molecules in the cube. In a red cube (on average) there are five times as many molecules in a one-wavelength cube. The percentage uncertainty is larger for the small blue cube because it contains fewer molecules. Therefore blue light is scattered more strongly than red: the skylight is a pale blue. An off-the-wall guess that the uncertainty in the blue cube is about 1 percent of the total gives a crude estimate of Avogadro's number; about a factor of 100 smaller than the best experimental value.

In 1910, Einstein published the exact answer to the blue-sky problem in *Annalen der Physic*, vol. 33, p. 1275. For those of us with limited proficiency in German, a summary and the resulting equations are on pages 102–103 of *Subtle Is the Lord* by Abraham Pais.

#17 Chlorophyll – Osmotic Pressure

When physical chemistry was young, just before the year 1900, osmosis was an important topic. Two water solutions, one dilute and one concentrated, were placed on either side of a membrane that could transmit only water. Water would try to cross the membrane into the more concentrated solution. If the concentrated salt solution is inside a closed, filled container, then pressure will build up because of the attempted water transfer. That pressure is called the *osmotic pressure*.

Today, cyanobacterial mats are typically found in flat coastal areas that are flooded only at the highest monthly tides. When I have visited them a few days after the highest tide, there were twinkly salt crystals (sodium chloride) on the mat from seawater evaporation. If a thundershower came by in the afternoon, the bacterial cells would go from saturated sodium chloride (26 percent salt, 5.3 moles per liter) to fresh rainwater within a few minutes.

The quantitative theory of osmosis was developed by van 't Hoff, winner of the first Nobel Prize in Chemistry. One version of his equation says that the pressure is given by $P = i\,C\,R\,T$, where P is the pressure, i is the number of ions from each mole, C is the concentration, R is the gas constant, and T is the thermodynamic temperature (formerly, the absolute temperature). A mole of sodium chloride dissolves into one Na^+ and one Cl^-, so i equals 2. It takes some fiddling to get the units correct, but the result is 1,328 pounds per square inch. However, evaporated seawater also contains magnesium, potassium, and sulfate, so the total osmotic pressure is around one ton per square inch.

Osmosis can be run backward. If you apply a pressure to the salty side, greater than the van 't Hoff equation indicates, then fresh water will be extracted from the salt water. This is called *reverse osmosis* and it is used commercially for desalinating seawater.

#26 Dolomite Twin Plane – Why It Is Impossible

Substituting a few percent magnesium for calcium makes calcite less stable. The hypothetical twinned dolomite would contain 50 percent magnesium. It takes about 3 percent of magnesium substituted into calcite to raise the calcite stability to the level of aragonite (Chave, Deffeyes, Weyl, Garrels, and Thompson, *Science* 173 [1962]: 33–34). The contrast in Gibbs free energy between calcite and aragonite is 1040 joules/mole. If it were linear, increasing the magnesium from 3 percent to 50 percent would correspond to 17,000 joules/mole.

A cubic millimeter of calcite (2.71 g/cm^3) with a molecular weight of 100 would consist of .0000271 moles. At 17,000 joules/mole, the energy required to twin a cubic millimeter of dolomite would be 0.46 joules. If the pocketknife blade moves through one millimeter, the force on the blade would have to be

460 newtons, about a 100-pound force. Before the force reached 100 pounds, you would either crush the corner of the dolomite crystal or cleave off a slice of the crystal.

#49 Penrose Diffraction – Source Code

Here's the 14-line BASIC program that computed the diffraction pattern in #49. I use the ancient FORTRAN custom of declaring all variables starting with $i – n$ as integers, and the rest as single-precision floating point numbers. Two 12,706 floating-point arrays, x and y, are the vertex locations in a Penrose tiling. The number .00005 in lines 2 and 4 is the wavelength (times some constants). The xx and yy loops – 2850 and 1950 – are the height and width of the book page in pixels. In line 2, the number 1425 is half of the page height and .21 avoids some unwanted symmetry if the picture centers on zero. In line 4, 900 is less than half of 1950 because part of the horizontal direction gets caught up in the book binding. (In publishing lingo, it bleeds into the gutter.) Again, the .46 part is a random small number to avoid introducing mirror symmetry.

The innermost loop computes the path length, in radians, from the source to the detector. This is Feynman's sum-over-histories. Discerning readers will notice that I moved the source and detector far away to avoid computing square roots to get the path length. Line 12 computes the squares of the summed sines and cosines. Notice that this is the diffracted energy. If we had wanted the amplitude, there would have been a square root in line 12. The constant .08 was adjusted to scale the energy to fit the 8-bit size of the .bmp graphics format. Not shown are the mechanics for writing the energy, z, into an output file.

```
FOR yy = 1 TO 2850
    dy = .00005 * (yy - 1425.21)
FOR xx = 1 TO 1950
    dx = .00005 * (xx - 900.46)
    sumc = 0
    sums = 0
FOR i = 1 TO 12706
    path = dx * x(i) + dy * y(i)
    sumc = sumc + COS(path)
    sums = sums + SIN(path)
NEXT i
    z = .08 * (sums ^ 2 + sumc ^ 2)
NEXT xx
NEXT yy
```

ACKNOWLEDGMENTS

Atomic Structure Databases

These databases describe atomic structures:

> Minerals: American Mineralogist: rruff.geo.arizona.edu/AMS/
> Inorganic structures: US Navy: cst-www.nrl.navy.mil/lattice/
> Zeolites: www.iza-structure.org/databases/
> Proteins: www.rcsb.org/pdb/ and also, www.wwdb.org
> Viruses: viperdb.scripps.edu

Most of the illustrations in this book were prepared using x-ray diffraction data downloaded from these databases.

Acknowledgments

Frank Truong, Department of Chemistry and Chemical Engineering, Cal Tech, for advice on biological structures

The photograph on page 45 was taken by Stephan Hoerold

The photograph on page 73 was taken by Tan Wei Ming

All other photographs were taken by Stephen Deffeyes